W0234667

Is Nuclear Power the Answer?

Is Nuclear Power the Answer?

John Blakemore

JENNY STANFORD
PUBLISHING

Published by

Jenny Stanford Publishing Pte. Ltd.
101 Thomson Road
#06-01, United Square
Singapore 307591

Email: editorial@jennystanford.com
Web: www.jennystanford.com

British Library Cataloguing-in-Publication Data
A catalogue record for this book is available from the British Library.

Is Nuclear Power the Answer?

ISBN 978-981-5129-69-4 (Hardcover)
ISBN 978-1-003-63657-1 (eBook)

About the Author

 John Blakemore is an accomplished pianist, sailor, consultant engineer, nuclear technologist, scientist, widely recognized polymath and an adjunct professor at the University of Newcastle, Australia. He has consulted to significant clients in Germany, France, Italy, Japan and China. He has invented and developed numerous business processes in a wide range of businesses, served on the boards of numerous companies, and has been an adviser to former prime ministers on innovation. He also invented a procedure in 1991 to save his eyesight. Dr Blakemore was the leader of the quality revolution in Australia with his book *The Quality Solution* and earlier led the first quality system in Australia in 1981.

Testimonials

"... I am astonished by your work. It is truly excellent."

Prof. Kenneth Preiss, Ben-Gurion University, Israel

"For my part, I am happy to recommend him, his research capabilities are evident, his energy is enormous and he is a first-rate experimental scientist."

Prof. E.O. Hall, University of Newcastle, Australia

"... his is an enviable reputation position among international writers."

Australian Institute of Management

"Dr Blakemore is an intelligent fanatic with very good professional skills and he is a good communicator."

Dr Ezzelino Leonardi, Technical Director, Pirelli

"... a pioneer for many decades in the effective use of technology in manufacturing and business ... a deep thinker."

Prof. Brett Ninness, University of Newcastle, Australia

"... blends deep concepts with prqctical examples."

Prof. Danny Samson, University of Melbourne, Australia

"... a great resource for our lean practitioners."

David B. Graff, United States Air Force, Air University, USA

"Dr Blakemore is recognised for innovation and invention ... he is recognised as one of the top engineers in Australia for manufacturing expertise."

Engineers Australia

Other Books by John Blakemore

1. *The Quality Solution* (1989)
   MASC Publishing
   Information Australia
   ISBN 0-7316-6907-X

2. *Quality Habits of Best Business Practice* (1995)
   Australian Institute of Management Competitive Edge Series
   MASC Publishing, Prentice Hall, and Pearson Education
   ISBN 0-13-242-835-0

3. *Strategic Planning for Business Self Help Model* (1998, 2011)
   ISBN 978-0-646-56012-0

4. *Velocity* (2011)
   MASC Publishing
   ISBN 978-0-646-55182-1

5. *Lean Six Sigma Service Management* (2011)
   MASC Publishing
   ISBN 978-0-646-57841-5

6. *Competitive Manufacturing Management* (2012)
   MASC Publishing
   ISBN 978-0-646-55182-1

7. *The Dating TrApp* (2018)
   ISBN 978-0-6483398-85 (Hardcover),
   978-0-6483398-9-2 (eBook)

8. *Technology for Business* (2023)
   Jenny Stanford Publishing, Singapore
   ISBN 978-981-4968-70-6 (Hardcover),
   978-1-003-38216-4 (eBook)

Contents

Preface

Nuclear power is anathema to some people but a saviour to others. People whose lifetimes have been extended and whose pain has been ameliorated with radiation therapy love it, but people cognizant of the terrible consequences of the nuclear accidents like Chernobyl and Fukushima hate it. In an ideal world, renewables like solar photovoltaics and wind turbines superficially offer the best solution. However, when these are critically examined, problems such as waste disposal, unreliability of supply, short lifetimes and the use of large areas of land cast a cloud over these, but with further development they could become part of the solution to supplying affordable reliable power. Nuclear power is not without its problems either, but it is reliable and the waste is small compared with renewables and is well controlled. Much of the criticism of nuclear power has been ill-informed.

This book touches on climate change and global warming and records the number of global warming periods that have occurred historically in Roman and Medieval times. It also points out that if the correct licensing and controls are in place and if the cost of global warming and the GDP flow are included, then nuclear is most probably cheaper than renewables.

As said in Introduction, this is a complex issue. It has been clouded by misinformation spread by ill-informed politicians and journalists. Nuclear power has been embraced by most of the developed countries in the world as it is cost-competitive with all other forms of energy production and generates no carbon dioxide while operating. A 1000 MW nuclear reactor can be built in 4 years, depending on local licensing and regulations, at a cost of $8000 per kW.

It is up to the government to develop the way forward using the best knowledge available and forget idealism. We must be pragmatic. The GDP flow-on effect has been ignored. If you become a more technologically informed society, benefits appear everywhere as the standard is raised. The most technologically advanced countries like the USA, Japan and France all embrace nuclear energy. We must join them. We are slowly recognising that we need help hence the AUKUS agreement for Australia to have nuclear submarines as part of its defence. We have recognised the medical needs as evidenced firstly, by Australia's first nuclear reactor, HIFAR (High Flux Australian Reactor) (10 MW), 1958–2007, which has been replaced by the OPAL (Open Pool Australian Light water reactor) (20 MW). The restrictions put in place in Australia to prohibit the use of nuclear power for energy generation are no longer valid and should be changed.

Australia has an opportunity to use the massive natural resource of uranium (we have 40% of the known world reserves), to generate a base load source of power for industry and domestic use and eliminate fossil fuel use. Base load power is a must for industrial development. Also, the GDP flow-on effect for the country will reap massive economic benefits and this is another major reason to embrace nuclear power as soon as possible.

Acknowledgements

The author acknowledges the numerous conversations of encouragement from a multitude of people professional scientists, engineers, chief executives, social workers and private friends. All these people have been thanked personally. The engineering professionals were all from the University of NSW, University of Newcastle, The University of Technology Sydney, and think tanks in Australia and one overseas. The private people are from the numerous friends in private industry and sporting organisations.

The author accepts full responsibility for the views expressed in this publication.

John Blakemore

Chapter 1

Introduction

It is hoped that this introduction will enable the reader to appreciate the complexity of the task ahead of them. In this book, we hope to provide the appropriate information to enable the general population to be more comfortable with nuclear energy as a viable alternative to generating power, particularly in the Australian context. It is written for the non-scientifically trained individual, but there are a few occasions when the use of mathematics is unavoidable, for example, when calculating the cost of electricity using the Levelized Cost of Electricity (LCOE) equation. Such interventions are minimised.

When the author looked at what was being said in the media, it was clear that there was a lot of confusion in the marketplace about nuclear energy. Much of this was fuelled by fear since the Chernobyl nuclear accident and later Fukushima were both horrific and both caused widespread destruction and radiation damage and, in the case of Chernobyl, loss of life. However, the public's major concern seemed to be nuclear proliferation and Hiroshima and Nagasaki and the bomb and the later H bomb.

Is Nuclear Power the Answer?
John Blakemore
Copyright © 2025 Jenny Stanford Publishing Pte. Ltd.
ISBN 978-981-5129-69-4 (Hardcover), 978-1-003-63657-1 (eBook)
www.jennystanford.com

From the author's standpoint, it was of great concern that there were numerous compromises made in the CSIRO GenCost (2023 and 2024) reports on asset life, cost of waste, and the cost of global warming. All these compromises placed nuclear energy at a significant disadvantage when the cost of electricity using the LCOE is determined.

Australia has an abundance of solar energy and wind as well as coal. It also has 40% of the world's known reserves of uranium. It is also a country with puzzling policies. We have a nuclear reactor—the OPAL at Lucas Heights, which replaced HIFAR, on which I was trained—but oppose it for power production. We are part of the AUKUS agreement, which will equip us with nuclear submarines which will not carry nuclear weapons. Political compromises by ill-informed politicians have hampered development.

It is typical that ill-informed groups of people will oppose change like the Greens did to try and stop the construction of the OPAL reactor.

As the land down under, we have been spared much but have contributed more than a little to assisting in solving world conflicts. In the eyes of many, this is regarded as somewhat puzzling as well as being amazing. Our role at Gallipoli in Turkey on 25 April 1915 and Villers-Bretonneux in France on 24–27 April 1918 is revered, whilst our predilection to fight wars for Britain and the USA is often found to be puzzling by people of most other countries.

Australia can be a modern utopia, just examine our enunciated goals. It is true that we have a good record in all of them:

- equality
- freedom of choice
- national sovereignty
- full employment
- conservation of the environment
- mature and balanced relations with all nations

This has been achieved with a continuously rising standard of living, social cohesion, a society capable of adapting to the rapidly changing world and a society participating in the decision making and directions of the nation. All this is achieved through a robust democratic system often but not always using vigorous debate. Open debate is the key to the future of the use of nuclear power.

All goals are lofty but reasonable. They can only be achieved, amongst other things, if Australia has a reliable infrastructure and a solid democracy.

We need a reliable source of energy. Historically this was supplied by coal. Will the future staple be uranium?

Everything we do requires energy, human or otherwise. To continue to raise our standard of living we need a reliable source of energy to power almost all aspects of our lives.

Nuclear power can do that. Can renewables like solar, wind, and tidal? We can argue that with a reliable storage system, and with an acceptance of the loss of huge amounts of land, sometimes arable, and good but yet to be developed, recycling systems for renewable waste, renewables could possibly do that too, but without a good storage system they cannot.

The Australian federal government in 2024 has a strong and unshakeable belief that the way to achieve a halt to global warming is to pursue a policy of creating all its power from renewable sources like solar and wind. We are not blessed with abundant hydro like many other lucky countries.

Renewables can play a part, but as this analysis will prove, nuclear power from uranium is more than cost-competitive with renewables. Fossil fuels are not regarded as renewable. Ideally, this would be wonderful if such a policy leads to a safe predictable power supply at the same time as the method of obtaining it does not contribute to global warming by producing CO_2 emissions or create a waste problem uncontrolled and the destruction of numerous valuable resources such as copper, nickel, cobalt, and in particular, rare earth materials and arable land.

Australia has 40% of the world's known reserves of uranium, so depletion of this element to drive a nuclear industry in Australia, is not a serious limitation. Nuclear power generation does produce, however, radioactive waste, and nuclear reactors as well produce plutonium, which can be used to make nuclear weapons. Plutonium can also be used to power fast breeder reactors, which in turn can produce even more energy but concerns about nuclear proliferation have meant that this option is not considered seriously.

Nuclear waste is stored and monitored, unlike waste from renewables. Also, for nuclear, there is not much of it, again, unlike renewables.

The plutonium produced has created a number of countries with nuclear weapons, so many that there is metastable equilibrium of the players in the world all of whom know that the use of nuclear weapons in another war could destroy the whole of mankind. Three times we have been on the brink: KE007, the Bay of Pigs invasion and a close call with an IT failure in Russia.

This book is written to unravel the complexities of the issues associated with nuclear energy so that we, as a society, can all make better and more informed decisions about its future and our future. It is written in the Australian context since Australia stands almost alone in the world with its unshakeable belief that renewable energy sources are the best solution to future energy needs.

This book offers an insight into the complexity of this issue from the perspective of a scientist with a PhD in nuclear physics, neutron bombardment (applied science) (1969), and post-doctoral qualifications in nuclear technology (Australian School of Nuclear Technology), and experience directly in the industry as a research scientist, who was part of a team trained to build the first nuclear reactor in Australia, planned for Jervis Bay in 1970. The author was a research scientist at the

Australian Atomic Energy Commission at the time and has operated a nuclear reactor, HIFAR (The Australian High Flux Australian Reactor), as part of his training. His role as a research scientist was to study the effect of neutrons on the mechanical properties of steels used in nuclear reactors. My PhD, awarded in 1969, was further developed extensively overseas to develop new higher strength steels, in particular, with only a small loss of ductility.

Consecutive Australian federal governments have continued to rule out nuclear power as a viable alternative energy source to satisfy Australia's current and future energy needs as they argue it is too costly. This book provides significant evidence to the contrary. The Australian federal government's preference is for solar and wind, renewables. They also rule out geothermal and tidal as well as fossil fuels.

Of course there are other arguments, danger associated with a meltdown, danger associated with radiation and danger associated with nuclear weapon proliferation are others, all reasonable concerns. All these are discussed, and fears should not be considered too seriously as the new generations of nuclear reactors are fail safe.

The federal Liberal government in 1970, under John Gorton as the prime minister, used an argument based on cost, to cancel the planned 500 MW(e) nuclear power station planned for construction at Jervis Bay. At that time, there were numerous very efficient coal-fired reactors in Australia, many built directly over coal seams in the Hunter Valley in New South Wales.

As mentioned earlier, the author was a research scientist at the Australian Atomic Energy Commission working on steel for nuclear reactors at the time. In 1970, global warming was not of much concern.

At a later date, in 1998, the Liberal government under Prime Minister John Howard confirmed that the course of action taken in 1970 was the correct one. However, he had no choice if OPAL was to be built. Hence, at the time it was a trade-off with the Labor Party and the Greens so Australia could build the OPAL (Open Pool Atomic Reactor) for medicine and research.

There was a strong recognition that Australia had to be capable of making isotopes for medicine and research as many of the isotopes needed had short lives and could not be imported from the USA or Japan.

Hence, Australia needed a nuclear capability to produce isotopes for disease treatment and prevention and for industrial diagnostic applications and research.

The present Australian federal government's policy position is somewhat perplexing since Australia has firstly its own nuclear reactor, the OPAL at Lucas Heights in Sydney, and intends via the AUKUS (Australia, United Kingdom, United States) defence agreement, to use nuclear-powered submarines in the future. These are not to be equipped with nuclear weapons. Japan and South Korea are both considering joining AUKUS as future partners.

To cap this off, we as a nation do not have a highly capable broad industrial and manufacturing sector as our manufacturing and industrial capability has gradually been dismantled since the end of the Second World War. This was done as tariffs were removed and the local manufacturing sector became too costly without that wall of protection.

During the period up to 1945, Australia built planes (the Wirraway), and later motor cars (the Holden), refrigerators (the Hallstrom), and most electrical home appliances as well as ships for commercial purposes and wartime. However, we no longer have this capability.

There is also some concern that many of the materials for solar cells in renewable solar farms, and batteries such as lithium, cobalt, nickel, and rare earths, may be in short supply in the future. No such limitation exists for nuclear as Australia has 40% of the world's known reserves of uranium.

Australia's OPAL nuclear reactor at Lucas Heights in Sydney is a 20 MW reactor using low enriched uranium and light water (ordinary water). This reactor replaced HIFAR (High Flux Australian Reactor), a 10 MW research reactor which was commissioned in 1958 and later decommissioned in 2007, a total of 49 years with no problems of any type. In fact, this reactor and its products are highly regarded overseas. Isotopes are produced by the OPAL reactor as for HIFAR, for Australia and the Pacific islands.

Small modular nuclear reactors (SMRs) offer a range of advantages. These have been in use successfully in ships and submarines for nearly 70 years. It is interesting to note that the sailors in nuclear submarines receive a radiation dose of approximately half that of people living in our capital cities, no doubt due to the effective shielding of the reactor vessel in the submarine itself, and the fact that the steel submarine body is also a very effective radiation shield.

OPAL, like HIFAR did beforehand, operates 300 days per year. As said earlier, the Greens Party tried to stop its construction and have opposed it at all times.

The author is an environmentalist at heart and wants a viable non-destructive solution to the world's ever-increasing energy needs.

Nuclear power generates very little CO_2 in operation and very little in the whole of the supply chain.

CO_2 is a colourless, odourless, non-poisonous gas, which is vital to all plant life and, therefore, animal life. It is an intrinsic part of the carbon cycle of nature. However, as a greenhouse gas, it is believed to be a major cause of global warming. It is not the only gas that contributes to the greenhouse effect. Even water vapour is a greenhouse gas.

The latest estimated cost of global warming is US$1.7 trillion per year but some estimates are as high as US$3 trillion per year. Fossil fuels such as coal oil and gas generate huge amounts of

CO_2; so, on this basis, their use must be minimised if CO_2 is the major cause of global warming.

There are 195 countries in the world, and 32 of them have nuclear power reactors. In addition, there are 413 nuclear power reactors currently operating, 61 under construction, 110 planned and 321 more are proposed. Clearly, most other countries believe that nuclear power is necessary to satisfy their own energy needs. Finland, the United Arab Emirates (UAE) and France have just brought new nuclear reactors on stream.

While excessive cost is used as the main argument against nuclear power, the second argument used is that it is dangerous because of the possibility of an accident and the danger of exposure of the population to radiation and death and the danger of nuclear weapons proliferation.

The argument that nuclear power generation produces plutonium, which can be used to produce nuclear weapons, has not been debated openly in Australia. Countries currently use their ability to develop nuclear weapons from plutonium as a deterrent to nuclear war, and fortunately so far this has worked.

The argument to shut fossil fuel power stations is on the basis that they produce excessive greenhouse gas emissions, mainly CO_2. This has received a lot of support as it is generally agreed that the CO_2 emissions increase global temperatures. However, there is also a strong view that global warming is the natural result of the long-term cycle of global climate change and the changes that occur are as a result of the variable radioactivity on the sun, including the approximate 10-year cycle, called the SOHO effect, and the 20-year cycle, and the variability of the elliptical path of the earth's orbit around the sun and its variable angle of tilt. Robert Carter's work clearly supports this.

There are other methods of energy production besides solar (PV and thermal) and wind, such as geothermal hydro and nuclear fusion and the use of thorium in reactors instead of uranium. Then of course there is the fast breeder reactor which is banned as it produces too much plutonium, so the worry with this is that nuclear weapons and terrorism would proliferate more quickly if we allowed everyone to build fast breeder reactors.

It is highly likely that the abandonment of nuclear power in 1970 probably retarded technological and industrial progress in Australia. Some evidence of the link has been published by the International Monetary Fund. In their publication, the Gross Domestic Product (GDP) impact per dollar spend is $4.31 for nuclear, $1.19 for renewables and $0.65 for fossil fuels. The original analysis is available Nicoletta et al. (2022).

So, therefore, the number of issues is significant and there are arguments for and against all the elements of importance.

Australia's heavy reliance on the mining industry and its conservative approach to technological innovation have now placed policy makers at a crossroads. Embrace nuclear technology and a new door to innovation will open.

Australia needs to overcome its political and cultural inertia; otherwise, this will continue to stifle technological creativity. We have been left at the starting gate.

This book, by design, provides a pathway to understanding the important issues raised and hopefully will enable the reader to be better informed and better participate and understand the political decision making.

The author has gone to great lengths to try and remove the politics, but sometimes this is not totally possible, and any bias is based on his personal beliefs; it is impossible to erase these completely as there have been many significant political decisions affecting the current status of the acceptance of nuclear power in Australian society.

Since this is a major issue and these issues are all fluid, the author would encourage readers to contact him to add to our understanding so as to benefit everyone.

John Blakemore
Consultant Engineer, Adjunct Professor
Phone: +61 (0) 414 970 758
mascjohn@icloud.com

Chapter 2

What Is Nuclear Energy?

God Gave us Uranium, Plutonium, Mendeleev, Einstein, Bohr, Heisenberg and Oppenheimer

Summary

Here is a condensed and simplified version of a quick course in nuclear physics without the mathematics. It attempts to give the reader a little background. It explains the difference between fission and fusion, the role of Einstein and Oppenheimer, quantum mechanics and the duality of light. It briefly explains how the nuclear reactor works and the types of nuclear reactors that have been built. Mother nature has built these also as nuclear reactors which have spontaneously occurred in nature. Waste control is discussed. The cost of building a nuclear reactor is also given. Using median data for the 400 nuclear reactors in the world, it is approximately $8000/kW.

2.1   Introduction

Uranium exists mainly in two forms in nature. uranium 238 (U-238) and uranium 235 (U-235). The numbers 235 and 238

Is Nuclear Power the Answer?
John Blakemore
Copyright © 2025 Jenny Stanford Publishing Pte. Ltd.
ISBN 978-981-5129-69-4 (Hardcover), 978-1-003-63657-1 (eBook)
www.jennystanford.com

refer to their respective atomic weights. As well, uranium has an atomic number of 92, and this is the position of uranium in the periodic table of elements. There are other forms as well, but these are not as important as U-238 and U-235. The most popular reactor types are briefly discussed.

Since this book may embrace a lot of new terms to the uninitiated, the author thought it would be advisable to introduce a few basic concepts to the reader as follows. So, we begin with an introduction to the magical and beautiful world of the atomic particles, the atom, the neutron, the electron and the proton. We do not go into the detail of quantum mechanics.

2.2   The Atom

The word atom is derived from the ancient Greek adjective, *atomos*, meaning "uncuttable" or "indivisible". The earliest concepts of the nature of the atom were debated in ancient India and ancient Greece.

The smallest constituent of all matter is known as the atom. Atoms consist of small positively charged protons and uncharged neutrons which reside in the nucleus. Negatively charged particles called electrons revolve around the nucleus to give the atom a neutral charge. The simplest element is hydrogen, and it has the lowest atomic mass and is given the atomic number of one as the first element of what is called the periodic table. It has one proton, one electron and zero neutrons.

2.3   The Periodic Table

The elements of nature can be arranged in a periodic table first postulated by Dmitri Mendeleev in 1889. This table of the 118 elements consists of periods and rows of elements which have the same or similar characteristics. It is illustrated in Fig. 2.1.

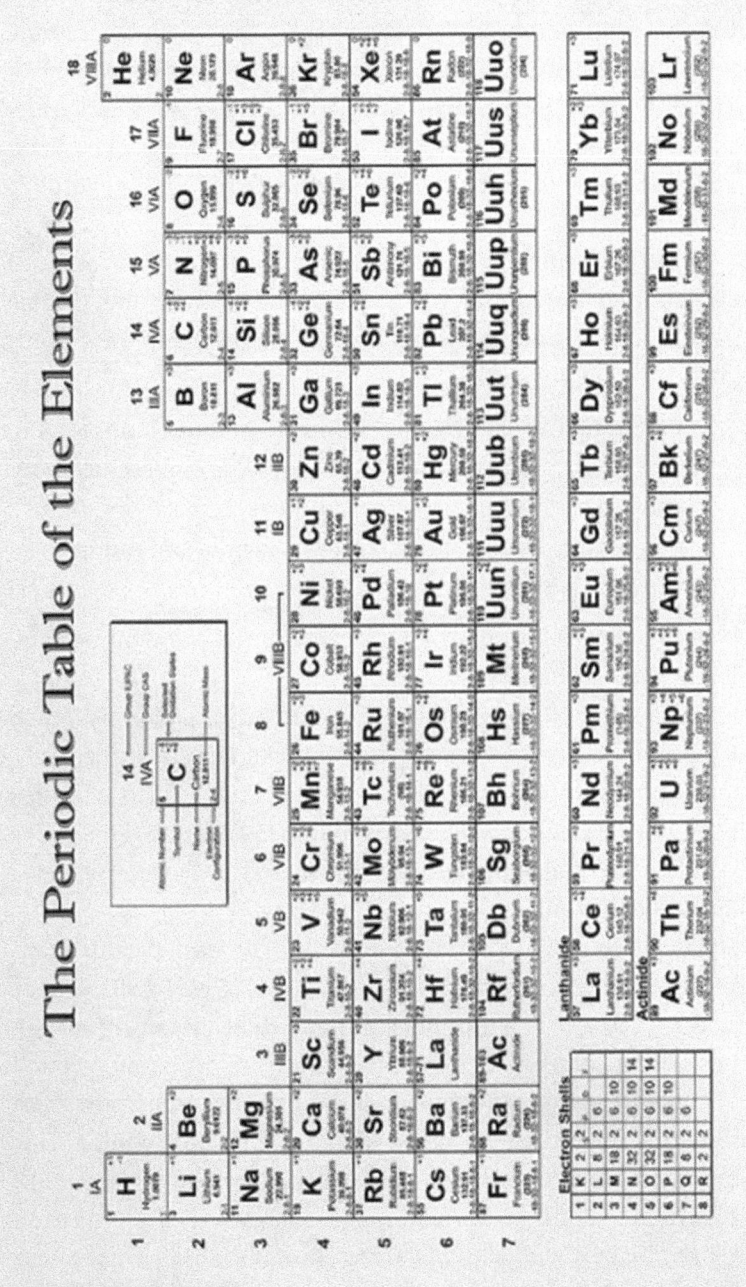

Figure 2.1 The Periodic Table of the 118 elements that appear in nature.

All elements are arranged in the order of their atomic number. The simplest element is hydrogen with an atomic number of 1 (AN = 1), and the highest AN is element 118 Oganesson, a noble gas, but could be a solid at 20° centigrade. Only a few atoms of this synthetic element have ever been made. It is named after the Russian scientist who discovered it in 2002.

Dmitri Mendeleev arranged the elements in vertical columns and horizontal rows as he had noticed the similar characteristics and performance of many elements both chemically and physically.

- Column 1a: hydrogen, lithium, sodium, potassium, rubidium, caesium, and francium—are similar in many ways.
- Column 1b: helium, neon, krypton, xenon, and radon—are all noble gases.

As we proceed from AN = 1 to AN = 118, the elements become increasingly unstable.

As mentioned, hydrogen is the simplest of all atoms. Much of the author's research was done introducing hydrogen into metals and alloys in what is called the first transition series of the periodic table. When introduced into metals and alloys by either thermal diffusion or electrolysis, hydrogen becomes a screened proton.

The author's target elements for hydrogen pickup were iron (Fe, AN = 26), cobalt (Co AN = 27), nickel (Ni AN = 28), copper (Cu AN = 29) and zinc (Zn AN = 30) and their mixtures (alloys). These were chosen because the electrons rotating around the nucleus are in energy bands and as we transition from nickel to copper the designated 3d band is gradually filled. This offered a very fertile area for discovery on how the mechanical properties of the material would change as the 3d band was gradually filled with electrons. In addition, hydrogen was becoming of increasing interest as a source of fuel and maybe the creation of advanced nuclear weapons and the hydrogen bomb which, as we will show later, may have become the next objective after the atomic bomb... Oppenheimer opposed this.

Hydrogen has two isotopes which are of significance in atomic energy, deuterium and tritium. Hydrogen has the atomic number of one, and has zero neutrons, one proton and one electron. Tritium has one proton, one neutron and one electron, whilst tritium has one proton, two neutrons, and one electron.

Like all the 118 elements in nature, uranium has unique properties. It has an atomic number of 92 and is part of a group of 14 elements called actinides. It has a number of isotopes, i.e. elements with the same atomic number, 92, and therefore the same number of protons but a different number of neutrons in the nucleus. There are 8 isotopes of uranium and the two most prevalent in nature are uranium 238, which is 99.2% of the uranium found in nature, and uranium 235, which is 0.7% of the uranium found in nature.

Uranium for nuclear reactors is most commonly "enriched". This means that the percentage of the uranium 235 element is increased. Therefore, there are more neutrons available after splitting the atom and this enhances the chain reaction producing energy. Natural uranium, uranium 238, can be used to generate power but to achieve reasonable efficiency, heavy water, deuterium water, must be used to control the population of neutrons available for the chain reaction and so avoid a meltdown and to generate power at reasonable efficiency. A moderator, usually carbon rods, are used to control the population of neutrons available for the chain reaction and so control the speed of the chain reaction.

To use natural uranium in the reactor as a fuel, it is necessary to use heavy water to achieve the desirable efficiency. Heavy water is extremely expensive and as a result, the natural uranium reactors such as the CANDU (Canadian Deuterium Natural Uranium) reactor and the SGHWR (Steam Generating Heavy Water Reactor) have not gained a lot of support as it is more economical to enrich the uranium than to make heavy water.

2.4   Energy Production

When uranium is bombarded with neutrons, the atom is split and a huge amount of heat energy is produced. This is illustrated in Fig. 2.2. This process is called fission.

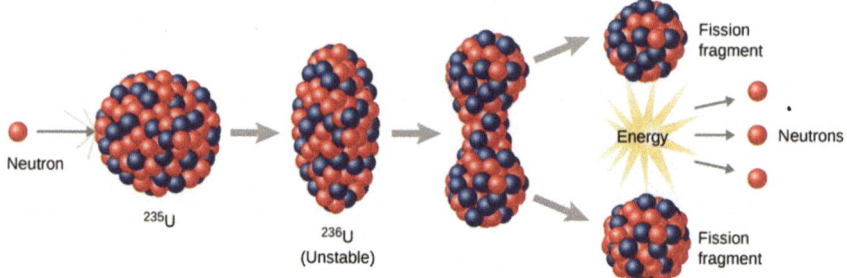

Figure 2.2 Schematic illustration of the splitting of the atom of uranium 235 when bombarded with neutrons. This sets up a chain reaction which can be controlled as it is in an atomic reactor, using a moderator to absorb neutrons. *Source*: https://opentextbc.ca/universityphysicsv3open stax/chapter/fission/. Reproduced under the terms of CC BY 4.0..

2.5   Einstein

Considered by many to be the greatest scientist of the twentieth century, Albert Einstein revolutionized scientific thought with new theories of space, time, mass, motion, and gravitation. Born in Ulm, Germany, in 1879, Einstein grew up in Munich, Germany. He originally failed in his bid to study at the Zurich Technical Institute. Unable to find a teaching job after graduating from a technical institute in Zurich, Switzerland, he accepted a post as an examiner in the Swiss patent office. He worked there from 1902 to 1909, devoting his spare time to his scientific interests.

In 1905, Einstein received his doctorate in physics from the University of Zurich at the age of 26 years and published three scientific papers, each of which had a profound effect not only on the field of physics but on the world.

His first paper explained the already-observed photoelectric effect—by which beams of light cause metals to release electrons which can be converted into electric current—by suggesting that light be thought of as discrete packets, or quanta, of energy particles.

For his work, Einstein, in 1921, received the Nobel Prize in Physics, but this was on the photoelectric effect, not atomic energy.

2.6 Energy = Mass times (Velocity of Light Squared) ($E = mc^2$)

The second paper, on the electrodynamics of moving bodies, put forward Einstein's special theory of relativity and contained the famous equation $E = mc^2$.

This equation showed that energy and matter are interchangeable. This provided the key to the development of atomic energy. (Here E is energy, m is mass and c is the speed of light.)

The speed of light, at 300,000 km per second (299,792,458 metres per second) is one of the few and perhaps the only real constant in nature.

The third paper virtually demonstrated the reality of atoms by showing that Brownian motion—the irregular movement of particles suspended in a liquid or a gas—is a consequence of molecular motion.

These papers earned Einstein professorships in Bern, Zurich and Prague. In 1914, he was appointed director of the Kaiser Wilhelm Institute for Physics in Berlin and offered a professorship at the University of Berlin; 2 years later, in 1916, he published his epochal paper on gravitational fields, "The Foundation of the General Theory of Relativity".

SOLVAY CONFERENCE 1927

A. Piccard, E. Henriot, P. Ehrenfest, E. Herzen, Th. de Donder, E. Schrödinger, J.E. Verschaffelt, W. Pauli, W. Heisenberg, R.H. Fowler, L. Brillouin
P. Debye, M. Knudsen, W.L. Bragg, H.A. Kramers, P.A.M. Dirac, A.H. Compton, L. de Broglie, M. Born, N. Bohr
I. Langmuir, M. Planck, M. Curie, H.A. Lorentz, A. Einstein, P. Langevin, Ch.-E. Guye, C.T.R. Wilson, O.W. Richardson

Figure 2.3 The Solvay Conference, 1927.

When Hitler and the Nazis came to power in Germany in 1933, Einstein, after a short period hiding in Scotland, emigrated to the United States, where he joined the newly formed Institute for Advanced Study at Princeton University in New Jersey.

Figure 2.3 shows a photograph from the Solvay Conference, a collection of the greatest scientists, engineers and mathematicians ever, here discussing atomic and quantum theory. This meeting and the discussions that followed added significantly to our understanding of the atom and sub-atomic particles and quantum physics. It is regarded as probably the greatest meeting of its type ever.

In 1942, after having been naturalized, Einstein was elected to full Academy membership and affiliated with the Academy's Physics Section.

After the Solvay conference, quantum theory began to gain many supporters. Einstein was not one of them. After all, he like most of us, had difficulty accepting that an atom or a particle can be in two places at the same time.

2.7 Quantum Mechanics and the Quantum Computer

With the invention of the Quantum Computer, it became clear that the switch which in a normal computer is either on or off, can, in a quantum computer be on and off at the same time. This is the reason for the incredible speed of the quantum computer. That is, a concept can be in two places at the same time.

2.8 Light

Light has a duality. It is both a wave and a particle. This is the foundation to the way we explain the double-slit experiment illustrated in Fig 2.4. This is an illustration of the light intensity if light is shone on a single slit and a double slit.

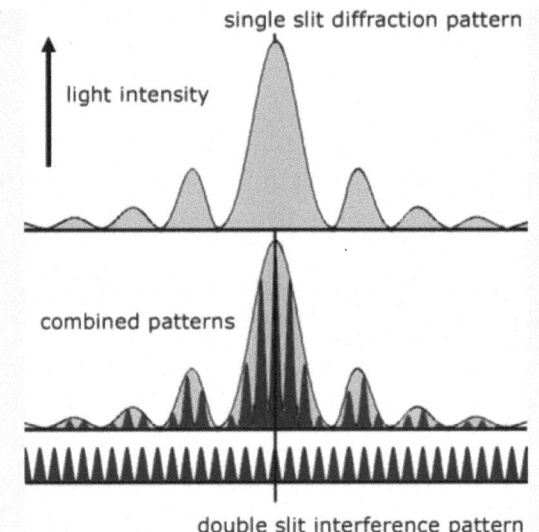

Figure 2.4 The double-slit experiment.

Einstein passed away in 1955, a lonely man who did not fully accept quantum theory.

At a later date, Richard Feynman, regarded as one of the greatest physicists of all time, said, "If you claim to understand quantum mechanics, then you clearly do not understand quantum mechanics."

After all, how can a substance be in two places at the same time?

This can be used to explain the duality of the wave theory of light.

The latest quantum computers use the fact that a point or switch can be positively charged and negatively charged at the same time. This is a result of quantum entanglement. To put it another way, a switch can be on and off at the same time. In conventional computers, the calculations are done with either a switch on or off but with quantum computers since the switch may on and off at the same time, then the calculation speeds are astronomically fast.

Figure 2.5 Einstein and Bohr in Copenhagen (left) and Einstein, a very lonely figure, in his later years, photographed at Princeton, USA (right).

Einstein would not accept quantum theory. He is believed to have said that "God does not play dice" alluding to the Heisenberg uncertainty principle and basic probability theory. Now we know from quantum mechanics that nothing is certain, and instead we can assign a probability density function to everything.

People today still grapple with the concepts of how a material or element can be in two places at the same time and how if you define its position then you cannot define its velocity and if you can define its velocity then you cannot specify its position.

2.9 Oppenheimer

Einstein was not a close friend of J. Robert Oppenheimer (1904–1967). Oppenheimer was an American theoretical physicist. During the Manhattan Project, Oppenheimer was director of the Los Alamos Laboratory and responsible for the research and design of an atomic bomb. He is often known as "the father of the atomic bomb".

By the time the Manhattan Project was launched in the fall of 1942, Oppenheimer was already considered an exceptional theoretical physicist and had become deeply involved in exploring the possibility of an atomic bomb. Throughout the previous year, he had been doing research on fast neutrons, calculating how much material might be needed to make a bomb and how efficient it might be.

Although Oppenheimer had little managerial experience and some troublesome past with associations with communist causes, General Leslie Groves recognized his exceptional scientific brilliance and so Oppenheimer was given the task to make the first atomic bomb. It was known that Hitler had been working on the bomb, but this was no longer the primary driver, as now the USA, after Pearl Harbour, had a new enemy: Japan.

Less than 3 years after Groves selected Oppenheimer to direct weapons development, the United States dropped two atomic bombs on Japan.

As director of the Los Alamos Laboratory, Oppenheimer proved to be an extraordinarily brilliant choice.

Oppenheimer was born on 22 April, 1904. His family was part of the Ethical Culture Society, an outgrowth of American Reform Judaism founded and led at the time by Dr Felix Adler. The progressive society placed an emphasis on social justice, civic responsibility, and secular humanism. Dr Adler also founded the Ethical Culture School, where Oppenheimer enrolled in September 1911. His academic prowess was apparent very early on, and by the age of 10, Oppenheimer was studying minerals, physics and chemistry. His correspondence with the New York Mineralogical Club was so advanced that the Society invited him to deliver a lecture—not realizing that Robert was a 12-year-old boy.

Figure 2.6 Oppenheimer with Einstein at Princeton University.

Oppenheimer graduated as valedictorian of his high school class, but he became ill with a near-fatal case of dysentery and was forced to postpone enrolling at Harvard. After being bedridden for months, his parents arranged for him to spend the summer of 1922 in New Mexico, a haven for health-seekers.

Robert stayed at a dude ranch 25 miles northeast of Santa Fe with high school teacher Herbert Smith as a companion and mentor. From there, he took five- or six-day horseback trips in the wilderness. This experience restored Oppenheimer's health and instilled a deep love for the desert high country.

Oppenheimer enrolled at Harvard in September 1922. He graduated in 3 years, excelling in a wide variety of subjects. Although he majored in chemistry, Oppenheimer eventually realized his true passion was the study of physics.

In 1925, Oppenheimer began his graduate work in physics at Cavendish Laboratory in Cambridge, England. J. J. Thomson, who had been awarded the 1906 Nobel Prize in Physics for detecting the electron, agreed to take on Oppenheimer as a student. At Cavendish, Oppenheimer realized that his talent was for theoretical, not experimental, physics, and he accepted an invitation from Max Born, director of the Institute of Theoretical Physics at the University of Göttingen, to study with him in Germany.

Oppenheimer had the good fortune to be in Europe during a pivotal time in the world of physics, as European physicists were then developing the ground-breaking theory of quantum mechanics.

Oppenheimer received his doctorate in 1927 at the age of 23 years and accepted professorships at the University of California, Berkeley, and the California Institute of Technology. At Berkeley, he became good friends with Ernest Lawrence, one of the world's top experimental physicists and the inventor of the cyclotron. Lawrence named his second son after Robert.

His security clearance was revoked in 1954 in a hearing during the Second Red Scare in the USA. Oppenheimer's old

Communist sympathies were dredged up and his clearance was revoked a mere 32 hours before it was set to expire. Oppenheimer had made political enemies by arguing against the development of the hydrogen bomb, and revoking his clearance stripped him of political power. The scientific community was outraged at the treatment of Oppenheimer.

Along with Albert Einstein, Bertrand Russell, and Joseph Rotblat, Oppenheimer established the World Academy of Art and Science in 1960. He continued lecturing around the world and was awarded the Enrico Fermi Award in 1963. He died of throat cancer in 1967 at the age of 63 years.

2.10   Source of Energy: Fission

A tremendous amount of energy is released when we split the uranium atom. Nuclear fission was discovered in 1939. This opened up a completely new source of power. Either uranium or thorium can be used to release this energy. These elements are both widely distributed in the earth's crust and Australia is fortunate to have 40% of the known reserves of uranium.

The energy released is in accordance with the Einstein equation, $E = mc^2$.

The isotopic composition of uranium's most important isotopes in natural uranium is given in Table 2.1.

Table 2.1 The isotopic composition of uranium's most important isotopes in natural uranium

Mass number	Weight percent
234 (U-234)	0.0058
235 (U-235)	0.711
238 (U-238)	99.28

A distinction must be made between fission and fusion.

2.11 Fusion

Fusion is the main atomic reaction that takes place in the sun and is responsible for all the light and energy we receive from it. The temperature of the sun is enormous, and therefore a lot of work on the fusion reactor in France, the ITER (International Thermal Energy Reactor), which has already cost the international consortium billions, tries to duplicate this at enormously high temperatures in a plasma which is contained in a magnetic field. Successful reaction times so far achieved are still only of the order of seconds.

2.11.1 The Fusion Reaction

Nuclear fusion is a reaction in which two or more atomic nuclei, usually deuterium and tritium, (isotopes of hydrogen) combine to form one or more different atomic nuclei and subatomic particles. The difference in mass between the reactants and products is manifested as either the release or absorption of energy.

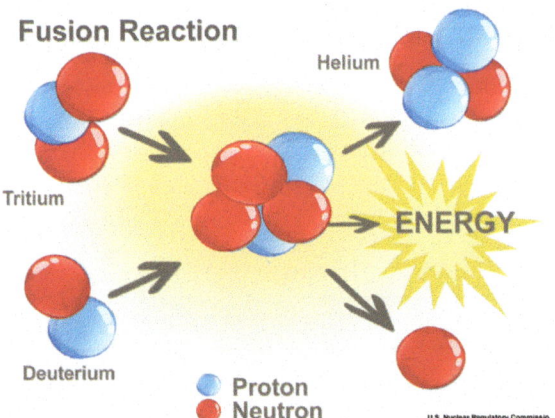

Figure 2.7 Nuclear fusion. This is the force behind the hydrogen bomb. So far, the attempts to build a sustainable nuclear fusion reactor have achieved a fusion reaction that lasts only seconds (the ITER reactor in France.)

2.12 Nuclear Fission: How a Fission Nuclear Power Reactor Works

There are numerous variations to the basics of a nuclear reactor, but all of them can be described in principle as follows.

There are six basic elements of a fission nuclear reactor:

- Containment building
- Reactor vessel, usually steel
- The reactor core
- The reflector
- Control rods (carbon or graphite)
- Heat exchanger

The energy and heat produced in splitting the atom is used to boil water which in turn is used to drive a turbine to generate electricity. These latter processes are the same for fossil fuel power stations.

As an engineer, it is clear, that the nuclear reactor is a very simple device. So simple that natural nuclear reactors have appeared spontaneously throughout history. It is simply a steel shell to contain the reactor core, fuel rods and control rods and externally a heat exchanger.

It is so simple that it is puzzling that some claims are made that it takes a long time to build a nuclear reactor. Why?

The answer lies in all the regulatory and licensing and quality control requirements imposed to ensure absolute safety because of the innate fear we all have of radiation and the atomic bomb.

2.13 Regulatory and Licensing of a Nuclear Reactor

Every component of a nuclear reactor is fingerprinted and 100% controlled and tested. This does not apply to solar, wind or other forms of energy production. All these controls are costly and time consuming. Many of these costs can be reduced.

However, why is the total time to build a nuclear reactor so important? It obviously increases cost, but this should not be a great concern as the cost is already factored into the cost of electricity in the LOCE (Levelized Cost of Electricity) equation which will be discussed later.

All the extra delays are man-made, and with a will and some innovation they can be reduced significantly. Japan, China, South Korea, Canada and France have all reduced the time to build a reactor substantially.

Also, multiple redundancy is built into the nuclear reactor. Whereas one gate valve would be satisfactory for a fossil fuel reactor, for a nuclear reactor there are five in series. This means that if one fails, then we have four more so that the probability of failure becomes almost zero, i.e. $(0.999)^5$.

Again, if the speed of construction is important, then we can make, say, four small modular reactors (SMRs) modules of 250 MW each and link them to form a 1000 MW reactor. These modules could be built in the factory and shipped to the site instead of building say a 1000 MW reactor on site.

2.14   The Steel Pressure Vessel

One of the author's roles in the nuclear industry was to study the effects of neutron bombardment on the steel in the pressure vessel. In 1969, pressurised water reactor (PWR) was a main contender for the planned 500 MW reactor to be built at Jervis Bay, New South Wales. This reactor had a wall thickness of 250 mm. Neither of the two major steel suppliers in Australia at the time, BHP and AIS, could supply steel of that thickness and of the correct quality. It would have to be imported.

In operation, the heat from the nuclear chain reaction is used to heat steam to drive a turbine. This part of the process is the same as that for a fossil fuel reactor. Hence as a fossil fuel station is retired, a small nuclear reactor can be readily installed and hooked up to the existing turbines and transmission system at quite a cost saving compared with building a solar farm and linking it with new poles and cables to the existing grid at some remote site.

2.15   Types of Nuclear Reactors Currently in Use in the World

Before discussing the types of uranium reactors, it is important to also introduce thorium reactors and fast breeder reactors, neither of which are being actively pursued like fission reactors and SMRs.

2.15.1   Thorium Reactors

There is three times as much thorium as uranium in the earth's crust. Thorium is not fissile, but it can be converted to a unique isotope, U-233 which is fissile. The simplest reactor would use a particle accelerator to produce the neutrons to bombard the U-233 to produce energy. These reactors are fail-safe, as to shut them down all we have to do is switch the particle accelerator off.

The waste produced by these reactors has a shorter half-life than those produced by uranium reactors.

Thorium reactors are still in the experimental stage.

2.15.2   Fast Breeder Reactors

Fast breeder reactors can produce as much plutonium from U-238 that could effectively multiply the uranium source by a factor of 50. They therefore generate more fissile material than they consume. These fast breeder reactors are expensive to build, but the real downside is that they produce large amounts of plutonium and the dangers of fissile materials falling into the hands of terrorists is compounded.

The world's last fast breeder reactor, the French Super-phénix, was shut down in 1998.

Research in the UK and France was abandoned because of the fear that to continue the risk of nuclear weapon proliferation was too great.

As well, a pro-nuclear MIT study expects that fast breeders are at least say 50 years in the future.

Number of operable nuclear power reactors worldwide

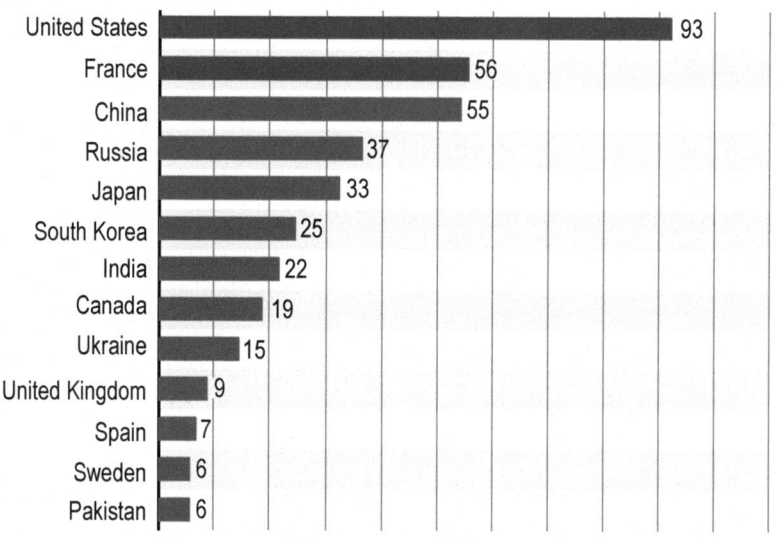

Figure 2.8 Number of nuclear power reactors worldwide.

The Australian Nuclear Science and Technology Association did produce a report claiming that a non-existent Power Station AP1000 could become competitive with coal under certain conditions.

2.15.3   Small Modular Nuclear Reactors

At the moment, small modular reactors (SMRs), are under development in 14 advanced countries throughout the world. These countries are as follows:

- USA
- Canada
- France
- Japan
- UK
- Denmark
- Sweden
- Czech Republic
- South Korea
- Russia
- China
- India
- South Africa
- Argentina

These small reactors are designed to be built in factories and assembled on-site. They are ideal for off-grid applications.

By not being involved at all, Australia will be left even further behind technologically. The rest of the world believes they should look at the nuclear option seriously as it is the only solution to reliable and cheap electricity and global warming.

We should be part of the technological and digital revolution and not simply hope the Australian mining industry will supply the necessary funds to support the Australian way of life as it currently does. We also need to use every opportunity to upgrade our engineering and technological capability so that if there is an emergency, we can cope without burdening our friends like the USA and then UK and maybe Japan, too much.

SMRs are currently in use in mainly maritime uses such as surface ships and submarines and in these applications their performance has been exemplary.

Small modular reactors offer significant benefits including the following:

- Lower cost
- Improved safety
- Versatility
- Waste reduction
- Ease of installation

2.16 SMR Lower Cost

Compared with large-scale nuclear reactors, the SMR construction time and construction costs are significantly less than that for larger nuclear reactors like a 1000 MW reactor and SMRs can be manufactured in a factory and shipped to the site.

2.17 SMR Improved Safety

SMRs have automatic fail-safe devices in addition to passive cooling systems and they require fewer mechanical parts and markedly less maintenance. They also have a proven record of successful safe use over 70 years.

The US navy has built over 200 SMRs with great success. They are an exceptionally safe form of energy production.

2.18 SMR Versatility

SMRs are factory built and can be readily transported. Their configurability is excellent, as in order to increase the output, you can simply add another unit.

2.19   SMR Waste

SMRs create less waste than large-scale nuclear reactors. New fuels are easier to recycle than in the past. An SMR could easily be the solution to our ageing coal-fired power stations since small units up to 300 MW can be retrofitted immediately into the existing power station.

2.20   The Versatility of SMRs

1. SMRs offer a great range of options, particularly in remote and off-grid areas.
2. There is a significant benefit economically from shorter manufacturing times.
3. SMRs can be easily scaled up.

Research and development of SMRs is continuing at a high pace, and it is likely that new classes of reactors will be born. This will likely include the recycling of existing used nuclear fuel and a further reduction in the amount of waste.

Currently, there are 50 SMRs under development worldwide.

Following are three of the most important developments:

- NuScale Model TM (PWR) using solid fuel with light water as a moderator.
- GE-Hitachi BWRX 300
- Terrestrial Energy Molten Salt Reactor

All these R&D programs are ambitious and demonstrate that the world outside Australia has a much more innovative and forward-thinking focus. They are all running massively over budget. This is not unusual when we are dealing with the unknown.

Nothing is certain when we are exploring and discovering.

2.21   Technology Is a Growing Organism

Technological advancement in the nuclear industry is growing exponentially.

With technology, as Alexander Fleming in 1928 found out, the final solution is not always obvious to some people, but it was obvious to him since he noticed that a substance within a mould spores of penicillin inhibited the growth of a culture. He immediately saw the opportunity and realised this could be used in treating infection.

Many other people like John Tyndall in 1876 and Andre Grassia in 1920s had noticed the effect before Fleming but did not foresee the medical use of it. Fleming saw the principle because he had been a doctor in the Great War and was appalled by losses due to battlefield infections and so he was open to see a purpose for what appeared to be a spurious effect.

Theories always start with general propositions and an individual technology or programme and an original idea to join the dots.

Technology is a very large part of what makes us human. The godhead resides quite as comfortably in the circuits of the digital computer or the gears of cycle transmission as he does at the top of the mountain or in the petals of a flower or the brain of a human.

We should not accept technology that deadens us, nor should we always equate what seems impossible, but what is desirable to what can or cannot be achieved.

We are human beings, and we need more than economic comfort, we need challenge we need meaning, we need purpose, we need alignment, we need nature, and this is where technology separates us from other beings.

Technology is our creation and that in turn creates more opportunities for us.

Nuclear technology, like all technology, affirms our humanness.

However, many of the benefits of SMRs are not available to Australia under current policy settings, so all Australia can do is look on and hope that we can influence the political decision makers.

2.22   History of SMR Development

In 1951, nuclear energy as a power source was born at 200 W.

In 1954 a 10 MW reactor propelled the USS Nautilus.

Fuel now in use is in the size of a thimbles. A thimble can contain as much energy as 5 barrels of oil (795 Litres.)

Whilst there is a strong focus on electricity production, this only produces 34% of our greenhouse gas emissions. Transport contributes 54%.

The NOAK (Nth Of A Kind), cost of an SMR is estimated to be approximately $5400 per kW.

The NuScale power module is 77 MW. Development in the NuScale design and like all Research and Development programs, there is a strong element of risk in achieving the desired objectives. The NuScale power module has already passed the safety standards with the US Nuclear Regulatory Commission (2020).

The GE Hitachi BWRX-300 is under development and is a boiling light water Reactor (BLW).

Terrestrial Energy is developing an Integral Molten Salt Reactor.

2.23   The Appropriateness of SMRs to Satisfy Australia's Needs

Australia with its vast sparsely populated land is a suitable candidate for the application and use of Small Nuclear Reactors (SMRs).

NuScale claim that 2 SMRs each of 77 MW could replace an ageing 150 MW coal- or gas-fired unit without the need for grid investment. This grid investment is often glossed over for renewables.

One of the significant advantages of SMRs is the reduced need for extensive grid investments, unlike renewables. SMRs can be easily integrated into existing networks.

It has been reported that there have been significant cost blowouts in the development of these new SMRs. Cost blowouts in R&D programs are not unusual. If we look at any infrastructure project in Australia or the USA, then cost blowouts appear to be what we all expect to happen when we are dealing with the unknown. Once these problems are solved and we enter the phase of commercialisation, then the cost figures become predictable and much less.

Often some journalists don't recognise this and report the cost blowouts in such a way that they appear to not understand the difference between the First Of A Kind (FOAK) and the Nth Of A Kind (NOAK) reactors.

This misleading interpretation unfortunately colours public opinion.

A useful analogy would be if it costs say $100 million to develop a new electric car then the cost of the first electric car may simplistically be $100 million. In fact, what we are saying here is that the FOAK may be $100 million but the NOAK will be many orders of magnitude less. This is normal and must be expected.

Australia, if it continues to follow a non-innovative path and waits for everyone else as we ride on the back of the mining industry, then it will never become a truly prosperous nation and will continue to be second rate technologically.

If I had not been prepared to be a guinea pig with my eyesight and risk being FOAK with the operation I underwent in 1991 to save my eyesight, then I would have been blind today like my mother was.

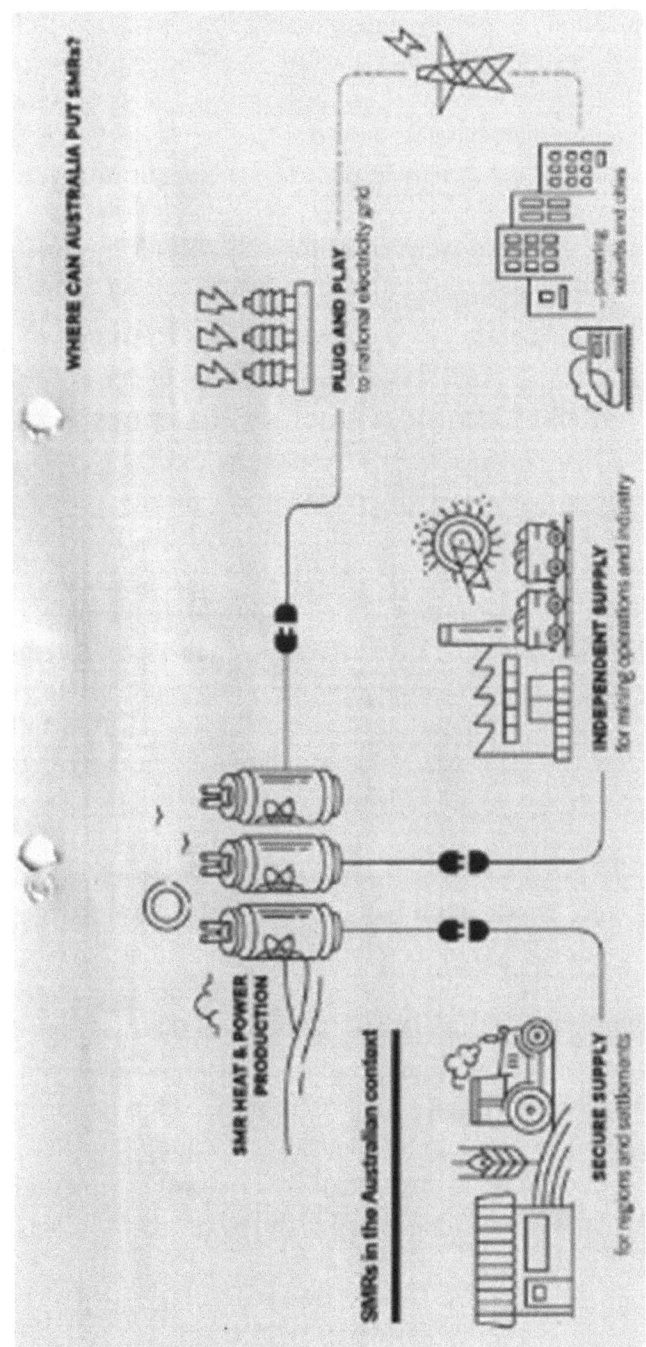

Figure 2.9 Small modular reactor (SMR) plug and play.

Innovation is not without risks, but we can't afford to let the rest of the world pass us by. This appears to be happening because we are resource rich and have successive governments that have the wrong focus. Maybe it is our culture and not just a characteristic of the Liberals or the Labor supporters or the Greens.

Nuclear reactors work by using the heat energy created from the splitting of the atom to heat water to generate steam to drive a turbine.

2.24   Most Common Nuclear Reactors

Almost all nuclear electricity currently is generated using two kinds of reactors: the boiling light water reactor (BLW) or the pressurised water reactor (PWR). These were developed in the 1950s, but they have improved significantly since then.

The first generation of reactors have most probably been retired now, but they could theoretically still be operating quite safely today. Most of the nuclear reactors being used in the world now are generation 2 or generation 3 nuclear reactors. New designs are coming forward all the time both large and small.

The fuel in the reactor is arranged in fuel assemblies in the reactor core. There might be up to 51,000 fuel rods in a core of a pressurised water reactor. The second most important part of the core of the reactor is the moderator material. The moderator's purpose is to slow down the neutrons and therefore control the chain reaction. The moderator can be what is called heavy water (water from the deuterium isotope of hydrogen) or graphite. During the operation of the reactor, some of the U-238 is changed to plutonium and plutonium 239 ends up as providing about 1/3 of the energy from the fuel in most reactors.

Advanced reactors are in operation in Japan, China, Russia and the United Arab Emirates (UAE), and others are under construction. Advanced reactors have enhanced safety and superior ease of operation. The latest reactors will burn actinides in the reactor, and these long-lived isotopes therefore do not become part of the spent fuel waste.

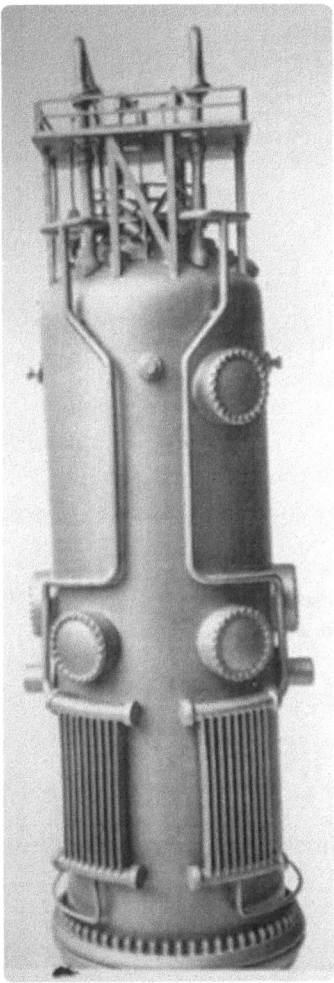

Figure 2.10 Small modular reactor.

Nuclear reactors can be created in a wide range of power sizes from 20 MW to larger than 1000 MW. The smaller ones, in particular, can be easily scaled up. Smaller ones used in ships and submarines are labelled SMRs or small modular reactors. These can be made in a factory and shipped to their ultimate destination readily. They can also be retrofitted to existing fossil fuel power stations and hooked up readily to the existing turbines. This flexibility gives them a big advantage over Solar and Wind since each of these would require a substantial investment in poles and wires to move the signal from their remote site.

Figure 2.11 SMR portability. Small modular reactors can be made in a factory and shipped to their destination. Modules up to 300 MW have been built. Over 200 have been built for naval use in ships and submarines. No problems have ever been reported.

2.25   Boiling Light Water Reactors and Pressurised Water Reactors

These reactors use enriched uranium as fuel and ordinary water (light water H_2O) as the moderator.

The Office of Nuclear Energy (2023) has produced the following schematic diagrams (Figs. 2.12 and 2.13) of the boiling light water (BLW) and the pressurised water reactor (PWR). These are the two most popular nuclear reactors in use worldwide.

PRESSURIZED WATER REACTOR (PWR)

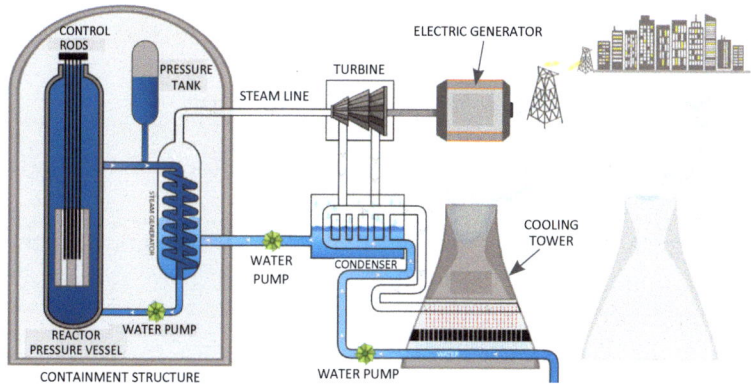

Figure 2.12 Pressurised water reactor.

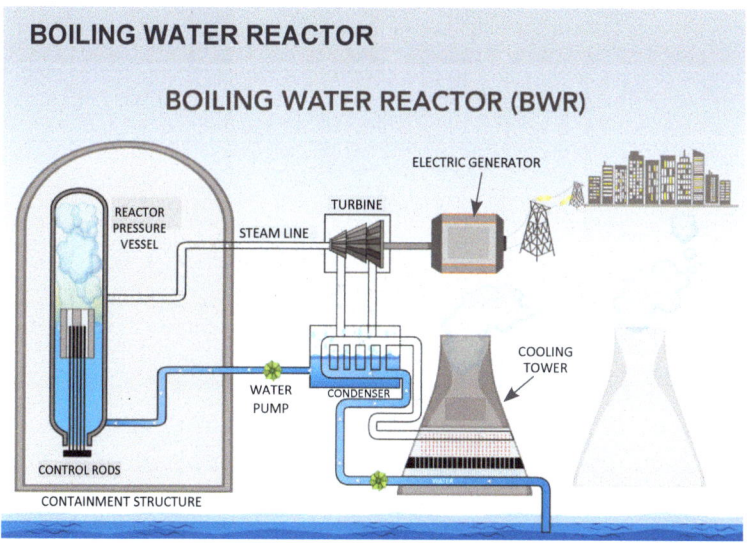

Figure 2.13 Boiling light water reactors.

2.25.1 The Pressurised Water Reactor

In a pressurised water reactor, steam is generated inside the large pressure vessel and fed directly to a turbine as shown in Fig. 2.12. These pressure vessels are made from steel up to 250 mm thick. Hence brittle fracture and fracture mechanics become of great significance.

2.25.2 The Boiling Light Water Reactor

In a boiling light water reactor, boiling water is fed directly from the reactor vessel.

Figure 2.14 shows the Canadian Natural Uranium Reactor Deuterium Reactor (CANDU).

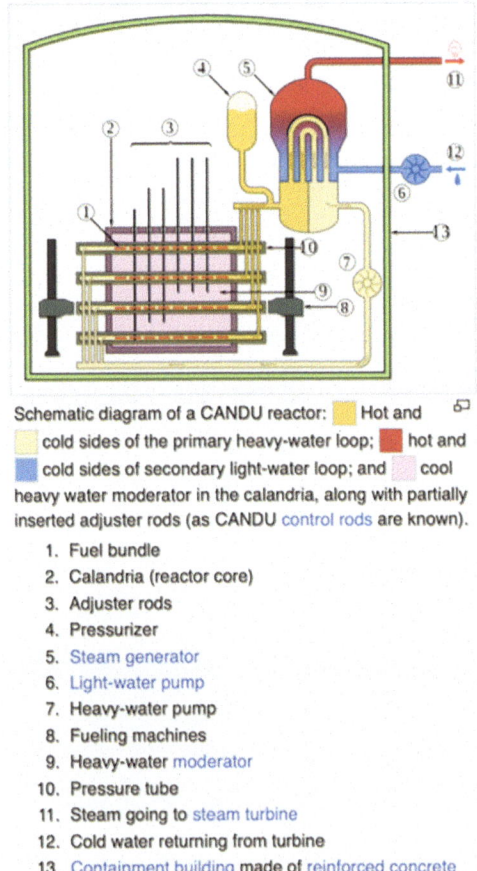

Schematic diagram of a CANDU reactor: Hot and cold sides of the primary heavy-water loop; hot and cold sides of secondary light-water loop; and cool heavy water moderator in the calandria, along with partially inserted adjuster rods (as CANDU control rods are known).

1. Fuel bundle
2. Calandria (reactor core)
3. Adjuster rods
4. Pressurizer
5. Steam generator
6. Light-water pump
7. Heavy-water pump
8. Fueling machines
9. Heavy-water moderator
10. Pressure tube
11. Steam going to steam turbine
12. Cold water returning from turbine
13. Containment building made of reinforced concrete

Figure 2.14 The CANDU Canadian deuterium natural uranium reactor.

Figure 2.15 shows the Open Pool Australian Reactor (OPAL).

Figure 2.15 The 20 MW OPAL reactor at Lucas Heights in Sydney. It is used to produce isotopes for medicine and research in Australia and the Pacific.

2.26 Waste Control

Unlike the renewables, the waste from a nuclear reactor is stored and controlled as it is radioactive. For solar and wind, 90% of the waste is buried in landfills after 15 to 20 years of operation. In a wind farm, the blades are made from fibreglass and the fibre and the resin used are toxic, and when buried they are not controlled.

Figure 2.16 shows a typical nuclear waste storage facility. It is above ground and tightly controlled and may become a source of energy for the future.

Nuclear waste is radioactive and therefore is controlled. It is not only a by-product of nuclear reactors but also a waste from hospitals and research facilities. Radioactive waste is also generated while decommissioning and dismantling nuclear reactors and other nuclear facilities.

Figure 2.16 Nuclear waste storage.

There are two broad classifications: high-level and low-level waste. Nuclear waste is processed to make it safe for disposal.

This includes its collection and sorting; reducing its volume and changing its chemical and physical composition, such as concentrating liquid waste; and finally, its conditioning so it is immobilized and packaged before storage and disposal.

The nuclear industry is the only large-scale energy-producing technology that takes full responsibility for the control of its waste and it fully costs this into the cost of producing electricity.

Solar and wind and other alternatives do not do this. On top of this, the amount of waste developed by nuclear power stations is relatively small compared with the other thermal electricity generation technologies such as solar and wind.

Nuclear waste could be treated not as waste simply but in fact as an energy resource. The waste itself from nuclear power stations is well controlled and can be managed very successfully in fact a lot better than the toxic waste from other services.

Safe methods for the final disposal of the high-value radioactive waste are technically proven. The latest-generation power stations, of course, will burn up the long-lived isotopes in the reactor themselves so they don't become part of waste stream.

Australia has developed the SYNROC or the ceramic rock system as a very sophisticated way to immobilise nuclear solid waste for final disposal to ensure that no significant environ-mental releases occur over tens of thousands of years.

Immobilisation of waste in an insoluble matrix such as a boron silicate glass or synthetic rock fuel pellets is another successful method. Waste is sealed inside a corrosion-resistant container such as stainless steel. To isolate waste from people and environment, it is often located deep underground in a stable rock structure, and this delays any significant migration of radionuclides from the repository where the containers are surrounded by an impermeable backfill such as bentonite clay or something similar.

2.27 Nuclear Reactors Have Spontaneously Occurred in Nature

A very significant case of naturally occurring uranium reactor, i.e. a reactor occurring spontaneously in nature, occurred about 2 billion years ago in what is known as Gabon in West Africa, where several spontaneous nuclear reactors operated within a rich vein of uranium ore. These natural reactors continued for about 500,000 years before dying away, and they produced all the radio nucleotides found in the high-level waste, including 5 tonnes of fission products and 1.5 tonnes of plutonium, all of which remains on the site, eventually decaying into non-radioactive elements.

Turbine blades from wind farms are buried in Wyoming, USA. Blades are made from fibreglass, which consists of fibre in a chemical resin. The fibre could cause as much damage as asbestos, and the resin could leach into the soil over time. This whole process is not controlled or monitored. Bloomberg has numerous photographs recording this, but these are not reproduced here because of copyright considerations.

2.28   Nuclear Power Costs

A recent publication by the Institute of Energy Economics (September 2024) concluded that household costs of power would increase if nuclear power was used. This study made a large number of assumptions, some of which are hard to justify. Much of the data was obtained from the CSIRO GenCost (2024) report. Let us look at some of them.

Efficiency 33%

I do not know how this figure was arrived at. It seems very low for nuclear but is probably reasonable for renewables like solar or wind.

Construction Time 3.3 years to 18.0 years

It is good to see some recognition of how factory manufacture and good planning and the application of modular reactors can reduce construction times for nuclear to as low as 3.3 years.

Waste costs

These are not included for renewables but are costed in for nuclear. If they are included for renewables, then the real cost of electricity would be greater. Renewable costs of burial and monitoring of the environment are ignored for renewables but included for nuclear.

Carbon dioxide emissions

This is a possible significant cost in the total supply chain but again is ignored. Wind turbine blades are fibreglass, last only a maximum of 20 years and release fine particulates as they age in operation, and the matrix will erode over time when buried. The emissions in operation of renewables and nuclear are similar, almost nil.

Capacity factor

Nuclear has capacity factor of approximately 90%, and for large-scale solar PV the factor is 32%.

Energy storage

This is not an issue for nuclear but is for renewables to cope with the intermittent wind and energy from the sun. Winds are highly variable, and the solar energy varies greatly. Therefore, storage is necessary. The most popular storage method is batteries, and this raises concerns about the depletion of natural resources like lithium and the materials in the new solid-state batteries. Uranium and thorium by contrast are in abundance.

Extra poles and wires

Solar voltaic and wind are in areas sometimes isolated and hence there is an additional cost of poles and wires. If small modular reactors are retrofitted to existing power stations, then this is not an extra cost. The cost for renewables is hard to quantify from the CSIRO GenCost data.

LCOE Equation

Using the Levelized Cost of Electricity (LCOE) equation with the correct costs over the lifetime of the reactors, renewables, and nuclear, and adding in possible waste costs, nuclear is probably cheaper than renewables. If the flow on technological advances is factored in, then the benefits to society are much greater indeed and decrease the relative cost of nuclear and make it an even better option.

2.29   Cost of Building Nuclear Reactors

Advanced nuclear reactors offer a new age in nuclear energy. Greater safety, improved efficiency and cheaper costs. The cost distribution for building advanced nuclear reactors is approximately an average of $8000 per kW, which equates to $8 million per MW, which equates to $8 billion per 1000 MW reactor.

The frequency distribution graph shown in Fig. 2.17 is a compilation of existing nuclear reactors worldwide. The median cost is approximately US$8000 per kW.

It takes approximately 6 to 8 years to build a nuclear reactor, but the Japanese have reported that they can do it in 4 to 5 years. Six to eight years is the average construction time globally. The Japanese regulatory system should be applauded and copied.

In Australia, it will take longer as Australia will have to import most materials, including the basic steel for the pressure vessel, as Australia's industrial capacity for a developed country is very poor.

Reactors can be built quickly but are often held back by stringent regulations and licensing and quality assurance requirements and fingerprinting. As mentioned, with the correct regulatory and legal and licensing arrangements, they can be built in 4 to 5 years once the correct regulatory requirements and licensing are in place.

Probability distribution of cost of advanced nuclear

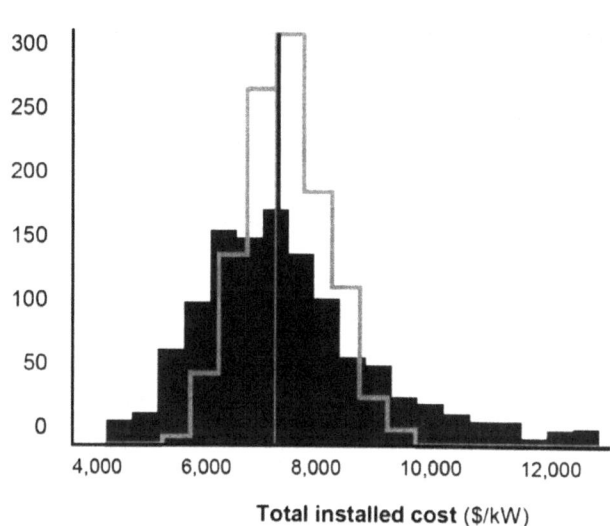

Total installed cost ($/kW)

Figure 2.17 Cost to build an advanced nuclear reactor.

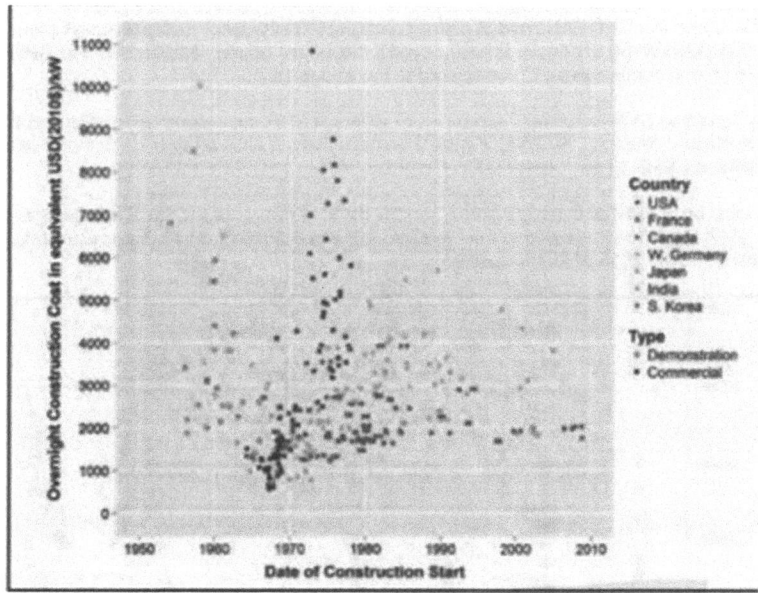

Figure 2.18 Distribution graph of the cost of building a nuclear reactor. Median is approximately US$3000 per kW.

Political restrictions and misinformation impede progress and often stop the progress completely.

The Greens tried to stop the construction of the OPAL reactor at Lucas Heights in Sydney. They were completely unaware of the need of the isotopes to treat cancer and provide the necessary isotopes for the gamma knife, sentinel node biopsies and other medical procedures to treat cancer to prolong life and reduce pain or even cure the disease in some cases.

Nuclear energy produces a large range of support for the medical industry and saves lives. Such interference by ill-informed groups adds to the delays and costs and endangers lives.

The average construction time varies from country to country. Japan, South Korea, and China are the fastest. The values quoted range from 3 years (Japan) to 20 years (USA).

2.30   Conclusions

This chapter discussed the basics of the atom and the history of the development of fission and briefly mentioned fusion. It also stated the latest understanding of the latest designs of nuclear reactors and their cost and time to build.

The rest of the world is embracing nuclear energy as the best solution to global warming and recognises that in the long term it may be cheaper than solar or wind farms.

The nuclear industry is very sophisticated and has proven to be very resilient and innovative.

Australia should embrace nuclear technology as the renewables cannot supply base load energy and the supply from renewables is largely dependent on the weather, but this can be partially solved with battery storage. Already the closure of the nuclear reactors and hence the removal of base load power has greatly increased power costs in Germany and is greatly adding to the inefficiency of the German industry.

Even with renewables with storage, there are problems with waste disposal and, in the future, possible material shortages and increased costs of cobalt, nickel, and rare earths and maybe lithium. There is no such shortage problem for uranium.

Economic Benefits of Nuclear Economy

Recent studies have concluded that the cumulated multipliers associated with the application of nuclear energy result in an approximate 4 times multiplier. For renewables, the multiplier is claimed to be approximately 1.2, and for fossil fuels it is approximately 0.5 (Nicoletta et al., 2022).

This finding is consistent with the fact that nuclear energy generates considerable employment and has up to a six times positive output effect. Also, the generation of power using nuclear power produces 25% more employment per unit of electricity than wind power. Nuclear energy spending multiplier is probably still larger than the renewable energy counterpart.

Figure 2.19 Chinese advances in nuclear energy. Chinese nuclear modular reactors ex-factory. Public Domain image.

China plans to build 150 new nuclear reactors between 2020 and 2035. Also in December 2023, China commenced operation of a fourth-generation gas-cooled nuclear plant in Shidaowan-1 (Shandong).

While Australia continues to ridicule the use of nuclear power, the rest of the world disagrees with our philosophy and is rapidly preparing for a new energy future.

There are many unsolved problems with renewables. Turbine blades are lasting less than 20 years and are being buried in landfills. Solar cells last a little longer but are also being buried in landfills. The particulate contamination of the air as the turbine blade ages and erodes is also of some concern. Nuclear waste is small by comparison with renewables and is stored and controlled.

The significant flow-on in the technological progress of the GDP growth of the nation using nuclear power is only now being appreciated. This is a factor of 4.5 for nuclear above renewables.

The Three Mile Island nuclear reactor is being refurbished at a cost of $1.6 billion and the chief beneficiary will be Microsoft as they gear up for the energy demands of artificial intelligence.

Once again, Australia will be left behind unless it recasts its vision and recognises that its focus on renewables without nuclear will be a failure.

Chapter 3

Climate Change

Climate change is real as is global warming.

How much is caused by carbon dioxide?

How much is natural due to the complexity of the interaction of the numerous variables not the least of which is the variability of the nuclear reactions occurring on the sun as it continues to heat up?

Summary

This chapter presents the background ideas which the world's best minds have on the causes and challenges of climate change. It emphasises that the problem is a very complex one and that even the experts have different views and whether the global warming we are experiencing is man-made (anthropogenic) or not.

No one can dispute the fact that the earth is fragile and heating up. However, can we change that trajectory?

Without the greenhouse effect, the earth would be inhabitable. The dangers of excessive heat are very serious. We must

Is Nuclear Power the Answer?
John Blakemore
Copyright © 2025 Jenny Stanford Publishing Pte. Ltd.
ISBN 978-981-5129-69-4 (Hardcover), 978-1-003-63657-1 (eBook)
www.jennystanford.com

learn to adapt. Wind power is a valuable resource, particularly for sailboats, but for power generation it has serious limitations.

Certainly, it can be readily demonstrated that there has been a significant increase in the average global temperature since the end of the Second World War (1945). The main culprits are believed to be carbon dioxide (CO_2), water vapour (H_2O), and methane (CH_4). These three are all natural and keep the earth significantly warmer than if there were no greenhouse gases present.

3.1   The Greenhouse Effect

Sunlight is radiated at visible and near ultraviolet wavelengths and this provides most of the energy incoming. After absorption, it is irradiated but at much longer, infrared wavelengths. Although the atmosphere is transparent to the incoming solar radiation, the irradiated energy from the earth's surface is strongly absorbed and reflected back by atmospheric water vapour and carbon dioxide so warming of the planet results. Without it the earth would not be habitable and the beautiful planet it is. On the earth, we live in a Goldilocks environment. The reason we can live on our planet at all is a result of this greenhouse effect.

This chapter will point out that the global warming resulting from the greenhouse effect is not only the result of a wide range of variables like radiation from the sun and other atmospheric gases and water vapour.

Each time the earth revolves around the sun, it traverses a different path. As well, the angle of tilt of the earth, which is responsible for our seasons, is slowly changing. Coupled with all of this is the fact that the sun is a star and the massive nuclear explosions on its surface vary dramatically all the time. As well, there is a 10- to 11-year cycle and then a 20-year cycle also, superimposed on this. A solar and heliospheric observatory (SOHO) has been set up to study the sun's behaviour.

There is still a strong body of belief that the global warming we are experiencing is part of long-term natural causes and the four warming periods that have been recorded are cited as proof that the global warming we are now experiencing is not man-made. Prominent supporters of this are Bob Carter (formerly Professor, James Cook University) and Professor Ian Plimer (University of South Australia).

As a scientist and engineer, the author's experience has always shown that as soon as you interfere with a system, then you change it. For example, if you wish to say measure the temperature of a liquid in a beaker with a mercury-in-glass thermometer by plunging the thermometer into the liquid, then the chances are that the thermometer, before it touches the liquid in the beaker, is at a different temperature to the liquid. Therefore, there will be heat transfer from one to the other. This will change the measured temperature of the liquid.

James Lovelock, regarded as one of the most important world visionaries on environmental issues, in his seminal book "Revenge of the Gaia", believes that CO_2 is the main culprit to global warming. He also believes that nuclear power is the best way to reduce the CO_2 in the atmosphere.

3.2 Our Beautiful Planet

Everyone knows that the earth is in bad shape. However, despite the rapid deterioration in the health of our planet, we live longer and healthier lives than ever before. Each time we are confronted with a major problem we, with our innovative approach, can generally solve it. Environmental examples are the reduction of lead (Pb) in our atmosphere and the closing of the hole in the ozone layer. There are many other success stories.

There is no doubt that the planet is heating up. Whilst most of the very intensive study has been for the last 300 years and the conclusion is clear that in that time frame the planet has heated up and this heating up is now accelerating. There is also a point of view that a 300-year time frame is too short to understand the underlying phenomena and the perceived heating is really part of a general variation in the earth's climate caused by a myriad of factors which are readily observable over longer time frames.

3.3   The Sun

The sun is the star of climate change. The sun shapes climate over timespans of thousands of years to millions of years to billions of years.

In 5 or 6 billion years, the sun will approach the end of its life. At that point, it will expand and heat up dramatically before cooling and becoming a white dwarf. It will become a massive super-hot planet before cooling to its white dwarf status. Its temperature is still on the heating-up trajectory.

Mathematician Milutin Milankovitch quantified the reasoning behind the changes in the orbit of the rotation of the earth about the sun and his theories have withstood vigorous debate for over 100 years.

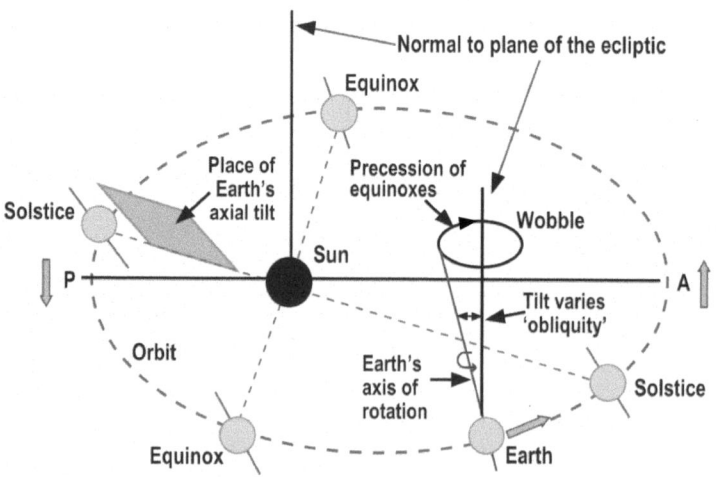

Figure 3.1 The earth's path around the Sun.

As mentioned, some of the variables determining how much of the sun's radiation affects the earth are as follows and illustrated in Fig. 3.1.

3.4 The Earth's Tilt:

- Minimum tilt: 21.8°
- Current tilt: 23.4°
- Maximum tilt: 24.4°

When the tilt is most pronounced, it allows for a stronger summer sun and a weaker winter sun. The tilt angle is gradually changing but will continue to increase to reach 24.4° and then start decreasing again. The main effect here is its influence on the relativity of the sun's radiation irradiating the hemispheres.

3.5 Greenhouse Gases

As mentioned, CO_2 is not the only gas believed to contribute to global warming.

The main greenhouse gases that do contribute are:

- Carbon dioxide (CO_2)
- Methane (CH_4)
- Water vapour (H_2O)
- Ozone (O_3)

CO_2 is produced when carbon burns and also when we breathe. Plants and the oceans absorb large quantities of CO_2 Plants require it to live.

Our whole biological life cycle is really a carbon cycle.

The prevalence of greenhouse gases in the atmosphere is as follows:

- Carbon dioxide: 380 parts per million (ppm) or 0.038%
- Methane: 1.8 ppm
- Ozone: 0.03 ppm

- Nitrous oxide: 0.03 ppm
- CFC HFC: approximately 1 ppm
- Water vapour: variable but up to 3% or more total in atmosphere

Svante Arrhenius predicted in 1908 that if the CO_2 doubled the earth's temperature would increase by approximately 5° Centigrade.

Charles Keeling used his experimental work at Hawaii's Mauna Loa to predict that the CO_2 will rise from 315 ppm in 1960 to 380 ppm of CO_2 by 2010. The famous Keeling Graph is given in Fig. 3.2.

THE BASICS

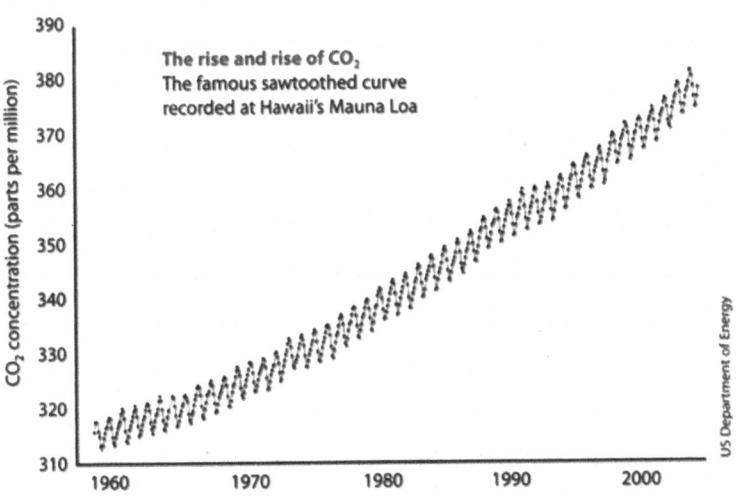

Figure 3.2 Keeling Graph showing the increase in CO_2 concentration in the atmosphere from 1960 to beyond 2000 as recorded at Hawaii's Mauna Loa.

The global atmosphere contains approximately 3000 gigatonnes of CO_2 (800 gigatonnes of carbon). The IPCC (International Panel on Climate Change), believes that the proportion of CO_2 produced by various sources is as follows:

- Industry: >40%
- Buildings: 31%
- Transportation: 22%
- Agriculture: 4%

3.6 CO_2 Production by Country

The Approximate Percentage of Global CO_2 produced by Country is given below:

- USA: 21%
- China: 15%
- Japan: 4%
- France: 1.5%
- Australia: 1.4%

Australia has a population of 27 million people and produces approximately as much CO_2 as France, which has a population of 65 million. France's much lower emissions of CO_2 per capita are a direct result of the fact that 75% of its energy comes from nuclear power stations and during power generation little or no CO_2 is emitted to the atmosphere, whilst in Australia 70% of the power is produced by burning fossil fuel such as coal or gas.

The only countries with emissions per capita greater than Australia are Qatar, the UAE, Kuwait, Saudi Arabia and Bahrain, all in the Middle East with a heavy reliance on oil.

Saudi Arabia has just commissioned nuclear power stations and so its emissions per capita will go down. The UAE has installed four 1400 MW plants (5.6GW(e)).

3.7   Dangers of Excessive Heat Caused by Global Warming

Heat wave conditions can produce death by the effects of the heat directly or due to particulate matter. Stagnant heat wave conditions can also create ozone. Pollution of the atmosphere is not only due to gases, but also, in part, due to particulates.

There are several studies that have linked deaths to ozone and particulates during the 2003 heat wave deaths.

3.8   Rain and Floods

Climate change can also increase the incidence of floods and catastrophic weather conditions.

Sea levels have risen 20 cm since 1880, and they continue to rise. There is a strong line of reasoning that the global warming we are experiencing is part of a long-term effect. This is illustrated in Fig. 3.3.

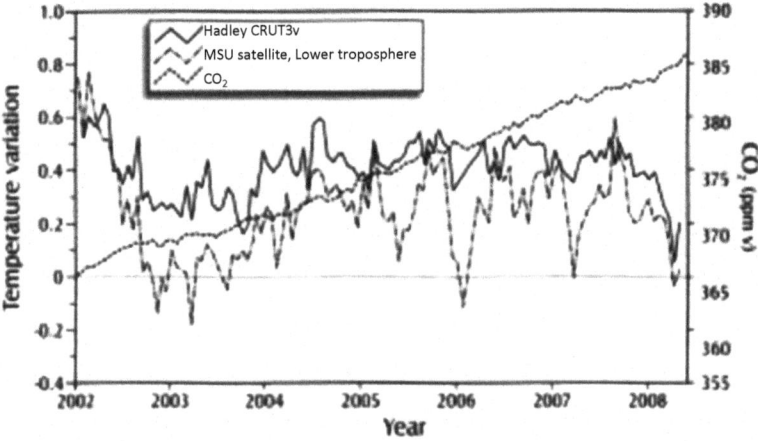

Figure 3.3 Relationship between CO_2 and global temperature variation from 2002 to 2008 (Plimer, 2009).

Scientists have studied a wide range of Proxy indicators of climate change including the following:

- Tree rings
- Lake sediments
- Coral growth rings

- Ice cores
- Ocean sediment
- Boreholes
- Old ground water
- Glacier moraines
- Sand dunes
- Coastal landforms
- Documented evidence

Figure 3.4 Iconic photograph of a polar bear on an iceberg in a sea of melted ice.

As mentioned, there is a strong case to say global warming is natural and was not as a direct result of CO_2 and the supporters of this view quote the following well-documented global periods.

1. The first of these was the Roman Warming Period from 250 BC to AD 450.

2. The Second was the Dark Ages from AD 535 to AD 900.

3. The third was the Medieval Warm Period from AD 900 to AD 1300.

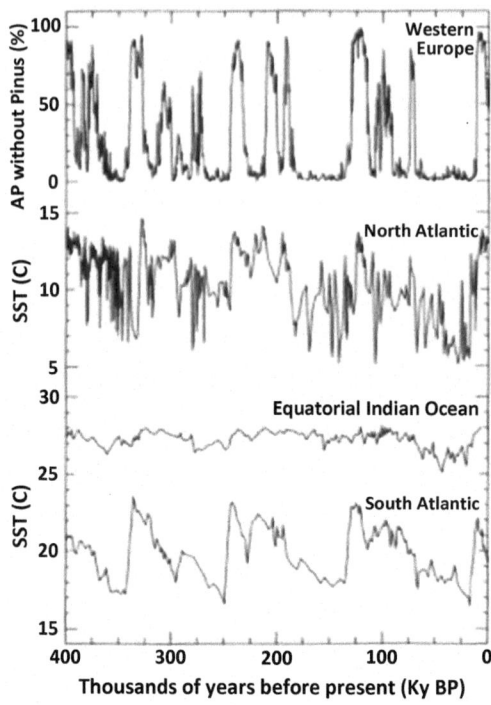

Figure 3.5 Warming periods in history. Climate variations over the last 400,000 years. SST is Sea Surface Temperature. Graphs from Fig. 2.23 in the IPCC 2001 report.

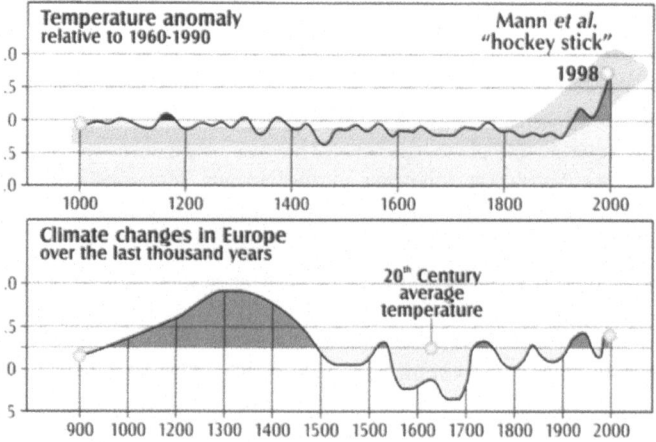

Figure 3.6 Climate changes AD 900 to 2000.

The little Ice Age followed the Medieval Warm Period from 1280 to 1850.

Then followed our current warming period, which started in the late 20th Century and continues today.

Whether the global warming we are experiencing is anthropogenic (caused by Humans) is only part of the issue and there is a view that while we try to stop the warming or at least slow it down, we are losing our focus on the bigger issues.

As a scientist, the author knows that the moment you intervene in a natural situation, you change the resulting output. So human intervention in natural processes must be avoided if possible.

It is believed that the variations in the solar energy on the earth cannot adequately explain the relative strength of the observed temperature warming.

Figure 3.7 Ancient tall ship. These would take more than 73 days to sail from England to Australia. Now catamarans can sail around the world in less than 41 days. (INDEC 3 Francis Joyon, 40 days 23 hours.)

Figure 3.8 Sail GP foiling catamaran. These can reach speeds of 3.67 times the true wind speed and sail directly into the apparent wind. Conventional keel yachts can reach speed of 0.8 times the true wind speed whilst planing dinghies can reach speeds of 1.5 times the true wind speed.

The concentration of CO_2 in the atmosphere at 378 ppm (0.0378%) is well above the pre-industrial level of 280 ppm (0.028%). As we attempt to decrease our emissions of CO_2, we hope we can slowly reduce the amount of CO_2 in the atmosphere and so slow down global warming. Meanwhile, the effect of global warming has already had a significant effect on natural ecosystems and human society.

The impact of global warming will become even more serious in the future. This means that we will have to adapt to the new conditions.

Chapter 4

Renewables: An Overview

Wind, solar (PV), solar (Thermal), and biomass are renewables, but fossil fuels such as coal and gas are regarded as non-renewables.

Summary

Australia is the world's largest coal exporter and is also the fourth largest coal producer in the world and the biggest emitter of carbon dioxide, the major greenhouse gas, per capita of any country in the world.

Reasonably efficient energy use and a wide range of renewable energy technologies are available and provided that we do the economic sums diligently, and as I show later in this book, we can prove that nuclear fission reactors are more than competitive with renewables on an economic basis and are superior otherwise in areas including storage and reducing global warming in particular.

Is Nuclear Power the Answer?
John Blakemore
Copyright © 2025 Jenny Stanford Publishing Pte. Ltd.
ISBN 978-981-5129-69-4 (Hardcover), 978-1-003-63657-1 (eBook)
www.jennystanford.com

4.1   Nuclear Energy and the Sun

The Sun is a huge nuclear reactor. The output of energy over a short period, varies gradually from time to time with dominant variations associated with an 11-year and a 22-year cycle variation. These are relatively small and their effects on the weather, I believe, are relatively small. More important than this, of course, daily are the variations in the earth's orbit around the sun, something not widely appreciated.

It is accepted that changes in the incoming solar radiation cannot explain any of the rapid warming observed in the last 20 to 30 years. It is well to keep in mind that to this day, no meteorologist has been successful in accurately predicting the weather beyond about 10 days, so one would argue how could they possibly predict what's going to happen in thousands of years. That is a reasonable argument. However, global warming is real and if you consider the time period to be long enough, millions of years, the earth will heat up as the sun continues to increase in temperature until it nears the end of its life and the sun becomes a white dwarf.

The melting of the Greenland ice cap is of great concern. The satellite study of this has indicated that the glaciers are carrying twice as much water as the North Atlantic Ocean. As a result, Greenland appears to be losing about 200 cubic kilometres of ice per year according to James Hansen, director of NASA Goddard Institute of Space Studies. Also, there are recent indications that positive feedback processes such as the building of the Arctic and Greenland ice caps and permafrost in two continents may be accelerating global climate change and global warming. Some of these processes may be irreversible at least on timescales for thousands of years. In fact, then one can conclude that the higher the increases in global average temperature, the higher the probability of irreversible changes.

The Stern Review was commissioned by the UK chancellor of the Exchequer and reported to by the chancellor and the Prime Minister. It was carried out by scientists in the government economic service of the United Kingdom and the former chief economist of the World Bank. According to the Nobel Prize economist Joseph Stiglitz, the Stern Review provides the most thorough and rigorous analysis to date of the economic costs and risks of climate change and the costs of risk of reducing emissions. It recognises that climate change presents a unique challenge for mankind. It is the greatest and widest ranging market failure ever seen to calculate the costs. The review uses an integrated assessment model that takes explicit account of the risks of various impacts, and it finds that "the total cost over the next two centuries of climate change associated under reasonable assumptions involves impacts and risks that are equivalent to an average reduction in global output capita consumption of at least 5% now and forever. While a cost estimate is already strikingly high, it also leaves out very much that is important.

Energy is the physical driving force of the universe and life and everything in it. It turns the wheels of industry, it transports us to school, work and holidays. It is incontrovertibly linked to mass through Einstein's famous $E = mc^2$.

Energy is classified as either kinetic or potential energy. Kinetic energy is energy of motion such as flowing water, wind, waves. Potential energy is energy stored in some form such as a lump of coal or in a battery or the carbohydrates and fats in food we eat.

In the international system of units, length is measured in metres and mass in kilograms. Walking at 6 kilometres an hour is equivalent to 350 W. Resting humans radiate 100 W.

Energy cannot be created or destroyed. It can only be transformed to another form of energy. This is called the first law of thermodynamics and is a fundamental law of physics.

The thermal efficiency of an energy system is defined as the useful energy output divided by the energy input.

4.2   Energy Needs

The approximate energy needs of the various major sectoral groups in Australia is as follows:

- 38% manufacturing construction transport
- 31% residential
- 30% commercial services
- 7% mining
- 6% agriculture
- 3% non-energy fuel
- 2% other

All renewable energy comes from natural sources, for example, solar from the sun bioenergy is solar energy stored in plants and organic materials or wind generated by uneven heating of the earth by the sun.

Waves, generally, are created by winds blowing on the oceans. Tsunamis are created by seismic shifts in the earth's crust, earthquakes. Hydroelectricity is likely powered by the sun which evaporates water from the earth's surface.

Tides are raised by gravitational forces of the moon and the sun. The intensity of solar energy input on the earth's surface is about 342 W per square metre. The primary energy consumption is about 0.03 W per square metre. Fossil fuels are defined as non-renewable although nature makes them from biomass over a period of millions of years.

It is generally very wasteful to produce hot water from electricity, whereas the amount of greenhouse gas produced by each sector depends on the source. Of the total CO_2 emissions, 35% are generated by electricity, 16.5% from methane and 13.5% from transport; these are the most significant sectors.

4.3   Emission Percentages by Sector

The energy related emissions from each commercial sector by energy service is as follows:

- Transport: 22%
- Lighting: 19%
- Air handling: 9%
- Heating: 4%
- Pumping: 22%

Emissions from the residential sector are as follows:

- Appliances: 42%
- Water heating: 27%
- Space heating and cooling: 18%
- Lighting: 8%
- Cooking: 5%

4.4   Wind Power

People in the past have used wind power for millennia in sailing ships and also to grind wheat and grain and pump water.

In the 1930s and 1940s, a new type of wind energy converter was developed that looked like a propeller and generated electricity for charging batteries on farms. These early wind generators, made in Australia by dawn light and typically had a maximum power output of up to 2 kW. The latest wind generators are grouped together in windy areas, but the output is highly variable due to massive fluctuations in the wind itself. Since they use a lot of land and are noisy, they are usually well away from people and hence the transmission costs can be high.

Large jumps in the price of oil particularly in the 1970s, sparked sales and a transformation of wind energy technologies that commenced in Denmark.

Figure 4.1 A wind Farm.

By December 2005, the installed capacity of wind turbines was 59,300 mW. Wind power is often regarded as one of the most environmentally damaging sources of electricity. This is because large areas of land are required for wind farms and the connections to existing grid from remote areas are not only expensive but also require a lot of land and transport. Since these are often installed on agricultural land, there is a loss of food generation capacity. It is estimated that 360 wind turbines need 200 square miles of land and 800 wind turbines equal 900 MW.

There has been some concern from various environmental groups that wind farms are dangerous because they kill a lot of birds. Statistically this probably does not have any great significance, but it, nevertheless, is environmentalists' concern. A bigger concern is the particulate matter shed by the blades as they gradually deteriorate and release these to the atmosphere.

The noise of the wind farm is often quoted as a common problem, but once again if we can move these far enough away from where people live, that issue is not terribly significant.

4.5   Noise Levels

Indicative perceived noise levels for various sources is as follows:

- The threshold of pain is 140 dBA.
- a jet aircraft at 250 metres 105 dBA
- pneumatic drill at 7 metres is 95 dBA
- city traffic 90 dBA
- a truck at 50 kilometres an hour or 100 metres 65 dBA
- car at 65 at 100 metres 55 dBA
- a busy road at 5 kilometres a wind turbine at 350 metres about 35 to 45 dBA.

The Australian Wind Energy Association claim that the general efficiency of wind farms is believed to be about 45%.

The general public regard wind farms as ugly. From the technical viewpoint, there are more serious problems. First of all, the energy supply is intermittent. The blades are in fact made from fibreglass, and currently these last only approximately 15 years. Because they cannot be recycled readily, the material is ending up in waste landfills.

It is important to recognise that we should not ignore the amount of carbon dioxide that is generated from waste. It is quite significant, and it is supposed to be or can be 3% of the total, so once again in the total cycle of wind farms this has to be taken into account as an extra cost. Since 2003, Denmark, has generated 20% of its electricity from wind power. This has allowed some coal-fired power stations to be retired in some areas, particularly places such as Denmark, where the wind blows fairly consistently over long periods of time and wind farms here are more economical than other countries where the where the wind velocity and strength varies dramatically.

In Australia, the capital cost of large installed wind farms is about AU$1800 per kW.

4.6 Biomass

Biomass could be a useful energy resource. It is material produced by photosynthesis or an organic by-product from a waste stream. It contains stored solar energy. It includes a wide variety of renewable organic materials, including forestry agricultural wastes and residues, urban trees, waste woody weeds, oil crops, plants, animal manure, sewage, dedicated energy crops and organic fraction of an organic fraction of municipal solid waste. In photosynthesis, growing plants capture solar energy water from the soil and carbon dioxide from the atmosphere and form carbohydrates and oxygen. The energy stored in the carbohydrates might be related by combustion of the solid fuel or by converting them into useful forms of stored energy. Biomass is converted into useful energy from the carbon dioxide that it originally absorbed from the atmosphere when it existed in the form of growing plants. Therefore, in theory, the use of biomass for energy production could be neutral in terms of net CO_2 emissions. The combustion of solid biomass is well understood, straight-forward, commercially available and readily integrated into existing infrastructure at low cost provided the fuel does not have that we transported long distances dedicated energy crops potentially. These crops include various kinds of trees, sugar cane, wheat, corn, sorghum, vegetable oil-bearing crops, sunflower, rapeseed oil, and soybean. Overseas, most of the agricultural crops are grown to produce liquid fuels such as ethanol from sugars and biodiesel from vegetable oil. In Brazil, millions of motor vehicles are fuelled mostly on 25% ethanol plus petrol with some 100% ethanol from sugar cane in the United States. Corn is the favourite crop for producing ethanol. Direct from solar is much more efficient and therefore would be encouraged if possible. So, matching the nature of supply with the demand needed is of critical importance to the overall efficiency of the system. Energy consumption is dominated by transport manufacturing and construction.

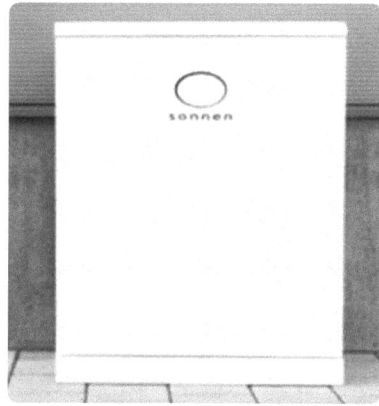

Figure 4.2 Compact home solar battery pack for storing extra solar energy.

4.7 Solar Energy Production (Photovoltaics)

Photons of sunlight are converted to electricity when electrons are produced by the solar panel. A typical efficiency rating of a solar panel is 16%. Sunlight is 1 kW per m^2, so a square metre of solar panels produces approximately 0.18 kW. Polycrystalline cells have achieved efficiencies of 20% and monocrystalline cells have produced efficiencies of 25% with a target of 30%. Cells commonly last up to 20 years, but their efficiency drops by up to 20%.

Solar offers many advantages. It is one of the few energy sources where the cost of the energy source is zero. In operation, however, it produces zero CO_2 but does produce significant CO_2 in the supply chain.

Panels are subject to the vagaries of the weather, dust, water, wind and high and low temperatures. There are claims that solar panels can last up to 30 years, but most data collected indicates a lifetime of 15 to 20 years.

The lack of supporting infrastructure is a major consideration since the panels are often installed in remote areas and require a substantial investment on poles and wires for transmission.

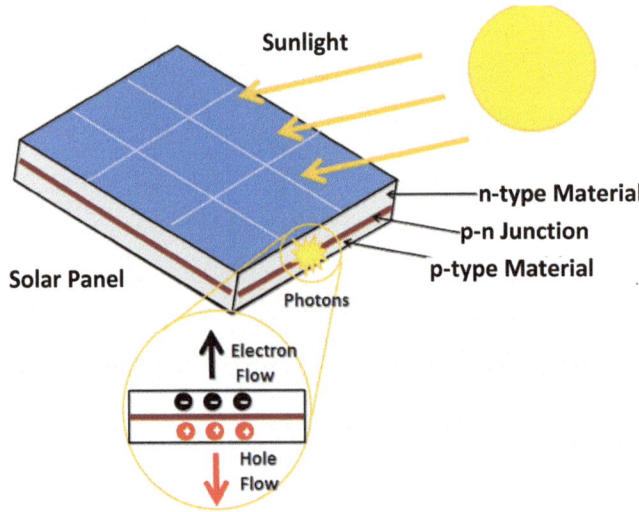

Figure 4.3 The principle of operation of the photovoltaic solar panel.

4.8   Perovskite Solar Cells

A huge amount of research is being done worldwide in developing more efficient solar cells. The promising of these appears to be the silicon-perovskite tandem cell. Efficiencies of up to 33.9% are being claimed from laboratory tests. Japan has already installed these cells on the walls of buildings and is planning a huge silicon-perovskite installation.

4.9   Snowy 2.0 Hydro

Snowy Hydro 2.0 is an energy storage system, not an energy generation system. The current cost so far is $13 Billion and rising. It started in 2019 and it is hoped to be complete by 2029. It is planned to produce 2.2 GW. It is called pumped hydro. Water is pumped from a lower reservoir to a higher one for storage and this water is released when needed. It is an energy storage system. It is not an energy generation system. It is useful for evening out demand loads.

Chapter 5

The Cost of Energy (Levelized Cost of Electricity [LCOE])

Summary

This chapter details the relative costs of nuclear power compared with renewables. It points out the limitations of the CSIRO GenCost (2023–2024) report and quotes the appropriate appendices in that report to back this up. Nuclear power is placed at a serious disadvantage in the way the parameter of economic life is inserted into the cost of electricity equation and the fact that the CSIRO ignores societal costs. Societal costs include the cost of waste and the cost of global warming. In all fairness to the CSIRO, the GenCost reports are only about the generation costs not societal costs.

5.1 The Cost of Electricity: Economics of Nuclear Power—The LCOE Equation

Nuclear power is cost competitive with other forms of electricity generation but offers the significant advantages of

Is Nuclear Power the Answer?
John Blakemore
Copyright © 2025 Jenny Stanford Publishing Pte. Ltd.
ISBN 978-981-5129-69-4 (Hardcover), 978-1-003-63657-1 (eBook)
www.jennystanford.com

base load reliability and zero CO_2 emissions. Fuel costs are a very minor part of the total cost of running a nuclear plant, but capital costs are in fact very significant compared with other alternative systems capital costs.

As well, coal- and gas-fired fuel generation costs are very much higher than intermittent renewables.

In assessing the economics of nuclear power, decommissioning and waste disposal costs are fully accounted for, unlike other generation methods.

Nuclear power plant construction is typical of all large infrastructure projects around the world where costs and delivery challenges tend to be underestimated. Capital costs include the cost of site preparation, construction, manufacturing commissioning and financing of the power plant.

Operating maintenance provision for funding and costs of decommissioning the plant and treating and disposing of used fuel and wastes should also be included, and for nuclear power they are, but for renewables this is not always the case.

The CSIRO GenCost (2023–2024), report gives the total energy need of Australia, currently, at 28,500 MW. This equates to approximately 10,000 kWh per person per year. Currently, 47% of this is satisfied by coal and a total of 68% is satisfied by fossil fuels as summarised in Table 5.1.

Table 5.1 Percentage of Australia's energy needs from each source of production

Method	Percentage
Coal	47.0
Gas	19.0
Solar	14.0
Wind	11.0
Hydro	6.0
Oil	2.0
Nuclear	0.0

5.2   Fossil Fuels

The burning of coal and fossil fuels produces CO_2 which the IPCC (Intergovernmental Panel on Climate Change) believes is the major reason the earth's temperature is increasing. Global warming is illustrated in the graph shown in Fig. 5.1. This graph is often referred to as the hockey stick graph. The rapid increase with industrialisation is clearly illustrated.

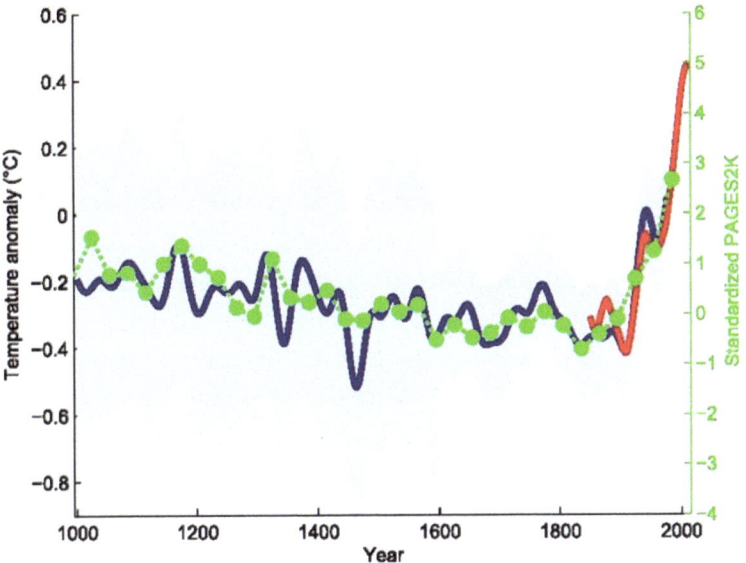

Figure 5.1 The hockey stick graph. *Source*: https://en.wikipedia.org/wiki/File:T_comp_61-90.pdf.

However, there are many other possible causes of an increase in global temperature as listed below and there are many eminent scientists who do not believe that CO_2 is the cause of global warming, and they cite data that indicates that the increase in CO_2 in the atmosphere followed the increase in global temperature not the other way around.

5.3   The Temperature of the Sun

As mentioned earlier, the sun is a radioactive ball of fire. The atomic reactions there are both fission and fusion and will continue unabated until the sun nears the end of its lifetime. It is believed that this mixture of fusion and fission will cause the temperature of the sun to continue to increase for the next hundreds of thousands of years, until it reaches a stage of decreasing radioactive explosions and it begins to cool before it dies and becomes a white dwarf.

5.4   The Cost of Electricity

The accepted method of calculating the cost of electricity is measured by a term called the Levelized Cost of Electricity, which is abbreviated to LCOE. It is a measure of the cost of electricity generation over the lifetime of the unit.

One of the key issues here is the lifetime of the unit (the asset life.) As you will soon see, this is a key issue in determining the cost of electricity. The nuclear asset lasts 60 to 70 years, while the solar or wind farm asset lasts only approximately 20 years.

The LCOE "represents the average revenue per unit of electricity generated that would be required to recover the cost of the building and operating a generating plant during an assumed financial life and duty cycle".

The method employed to calculate the cost of electricity is given below. It is not necessary for the reader to fully understand the maths. The parameter inserted are critical in determining the outcome. The CSIRO uses economic life based on an assumed finance of 30 years for solar, wind and nuclear. The asset life of nuclear is 60 to 70 years and that of solar and wind is 20 years, so this assumption favours the renewables. Australia's HIFAR reactor was decommissioned after 49 years as it was too small to meet the demand. In the USA, the government will finance the nuclear reactor and this will be the case probably in Australia in the future. This will mean the comparisons can be made on the real asset life and this increases the cost of electricity from renewables and decreases the cost for nuclear.

All this may not be immediately clear, but for those who wish to challenge this assertion, the relevant equation is given below for the LCOE.

$$\text{LCOE} = \frac{(\text{Sum of costs over lifetime})}{(\text{Sum of electrical energy produced over the lifetime})}$$

Here,

$$\text{Sum of cost over lifetime} = \sum_{t=1}^{n} \frac{(I+M+F)}{(1+r)^t}$$

and

$$\text{Sum of electrical energy produced over lifetime} = \sum_{t=1}^{n} \frac{E}{(1+r)^t}$$

where

I = investment expenditure in year t

M = operations and maintenance in year t

F = fuel expenditure in year t

E = electrical energy generated in year t

r = discount rate

n = expected lifetime of the reactor

The CSIRO GenCost (2023–2024) report uses lifetime values for solar photovoltaic farms and windfarms as 30 years and for nuclear reactors as 30 years.

These values are possible outliers, and median values would be more reasonable at say 60 years to maybe 70 years for nuclear and 15 to 20 years for solar PV and wind with a possible extreme life of 25 years.

If we compare the capital costs ($/kW) and the LCOE ($/MWh) for small modular reactors from overseas data and compare it with the CSIRO GenCost figures, the differences can only be explained by the CSIRO using outlier data in the LCOE equation. It is unreasonable to use such selected data.

Using outlier data in the Intelligence Quotient (IQ) assessment would say all people have an IQ of 160, i.e. the same as Einstein. The CSIRO has used parameters that favour renewables and penalise nuclear. If median data is used and the cost of capital is discounted over the correct values of the lifetimes of the various methods of energy creation, nuclear would possibly be the cheapest particularly when the cost of global warming is allowed for.

The discount rate assumed by the CSIRO GenCost (2023–2024) report is 0.0599%, which is reasonable.

5.5 SMR Costs from the CSIRO GenCost and Overseas Values

If we factor median values as per the work in Canada and the USA, the values above for the SMRs it is clear from the table below that there are massive differences in the calculated costs. The GenCost results are as a result of the assumption that the nuclear reactor will last only 30 years. The capital should be discounted over the true lifetime of the asset.

		Capital ($/kW)	Capacity factor (%)	LCOE ($/MWh)
Small modular reactor	Low	4,837	90	66
	Median	7,098	90	87
	High	16,487	90	117
GenCost (CSIRO 2023)	Low	7,237	80	143
	High	16,487	60	336

The discrepancies between the LCOEs quoted for the SMRs here are huge and of some concern.

The author has expressed his views to the lead researcher and author of the GenCost (2023) report,

Dr Paul Graham, and he has replied by saying that other people have also expressed similar views to mine expressed here.

GenCost is continually updated and has a very robust communication system. As a result of these concerns, CSIRO has included significant concessions as detailed in their appendix in GenCost (2023–24) report.

5.6 Appendices of Significance in GenCost (2023–2024) Report

Appendix D.4.1 (page 107)

It has been suggested that nuclear could access special financial arrangements to move the economic life closer to the asset life. However, CSIRO prefers to not introduce special arrangements for technologies where there is limited Australian evidence. In the USA, it is common for the economic life of the reactor to be asset life. In Australia, this problem could be immediately solved by the government owning the nuclear reactor. This attitude and policy of the CSIRO places the nuclear reactor costs at a serious disadvantage.

Appendix D.4.17 (page 117)

The total cost for electricity should be the system cost. However, CSIRO dismisses this as follows:

> *The stated purpose of GenCost is to provide essential capital cost information for the modelling community to use in their own studies.*

Put another way, CSIRO does not include the cost of global warming into the LCOE equation so once again nuclear power is disadvantaged.

5.7 Discussion

The largest single factor responsible for the significant differences in cost is the assumed lifetime of the reactor, t, as it appears in the discount rate when it used in the mathematical expression, for the lifetime of the reactor as follows $(1 + r)^t$. From the following table, the sensitivity of the discount factor to lifetime is clear and as a result the calculation favours solar and wind if the lifetime of the asset is assumed to be 30 years for both. Using 20 for solar and wind and 70 for nuclear paints a very different picture remembering that this factor appears in the denominator of the LCOE equation.

$t = n$ (lifetime)	$(1 + r)$ to the power n
1	1.06
2	1.12
10	1.79
30	5.73
60	32.8

From this table, the sensitivity of the calculated LCOE to the lifetime of the reactor, n, is readily apparent, and since it is the denominator of the equation, then the effect is very substantial.

Also, the LCOE equation has limitations. It does not include the cost of the waste but the technology roadmap for small modular reactors does. The waste from Solar farms and wind farms is being buried in landfills. Again, the CSIRO GenCost figures and the LCOE equation do not include the cost of global warming estimated to be between US$1.7 trillion to US$3.0 trillion per year. The exclusion of this cost significantly favours renewables and places nuclear at a sizeable disadvantage.

The inputs to the LCOE calculation do include the following:

- Cost of capital
- Decommissioning
- Fuel costs
- Fixed and variable operations and maintenance costs
- Financing costs
- Assumed utilisation rates

$$\text{LCOE} = \frac{\text{(Sum of costs over lifetime)}}{\text{(Sum of electrical energy produced over the lifetime)}}$$

GenCost Data (2023–2024)

x		GenCost Solar PV	GenCost Nuclear
I	Investment expenditure year 1($/kW)	41	345
M	Operations and maintenance year 1	17	205
F	Fuel expenditure year 1	0.0	0.5
r	Discount rate	0.0599	0.0599
N	Expected lifetime of station (years)	30	30

The assumed GenCost life of the nuclear reactor of 30 years places the nuclear reactor at a serious disadvantage as far as cost is concerned. An assumed lifetime of 70 years or more is more reasonable.

5.8 Special Notes on Table Above

The cost of waste is not included for renewables but is included for nuclear.

The 30-year lifetime for solar is well in excess of the values obtained so far. A value of 15 to 20 years is more realistic.

The 30-year lifetime assumed for nuclear is much less than what is experienced overseas. A value of 60 to 70 years is more realistic.

These two variations greatly favour solar and place nuclear at a disadvantage.

1. Output costs per unit of energy ($/MWh), already account for the capacity factor, while output costs per unit power ($/MW) do not.

2. The cost of capital is a contentious issue. Discount factors can range from 6% to 16%. Assuming a low discount rate favours nuclear.

3. LCOE does not include the social cost of greenhouse gas emissions and this puts nuclear at a disadvantage.

4. The LCOE is inversely proportional to the size of the unit.

5. The HIFAR reactor in Australia lasted approximately 49 years and was still operational when decommissioned. It needed upgrading from 10 MW to 20 MW.

The significant differences in the parameters between the GenCost data for nuclear SMR and overseas data, particularly the SMR Roadmap, leads to the following discrepancies in LCOE shown in the following table.

Source	Capital ($/kW)	Capacity (%)	LCOE ($/MWh)
GenCost (low)	7237	80%	143
GenCost (high)	16487	60%	336
SMR roadmap (low)	4837	90%	66
SMR roadmap (median)	7098	90%	87
SMR roadmap (high)	9476	90%	117

The differences between the LCOE values are very high.

They come about because of the big differences in the assumed life of the solar PV and nuclear and the capital cost differences.

The LCOE for Solar PV claimed by the CSIRO is $81/MWh ($8100/kWh). So realistically, from the above, SMRs are competitive on a cost basis even when handicapped by not including the full asset life in the LCOE equation and excluding the cost of renewable waste from the cost of solar or wind.

5.9   My Letter to CSIRO Lead Author, Dr Paul Graham, Regarding the Cost of Electricity

Dear Dr Paul Graham,

It is my belief that you have inserted outlier parameters for the lifetime of Solar PV and Nuclear that give an incorrect view of the true relative cost of electricity over the lifetime of the solar station and the nuclear station. Therefore, please see my assessment below:

The Cost of Electricity from the accepted Levelized Cost of Electricity equation. (LCOE).

LCOE equation = $\sum_{t=1}^{n} \dfrac{(I+M+F)}{(1+r)^t}$ = *sum of cost over lifetime*

Divided by

$\sum_{t=1}^{n} \dfrac{E}{(1+r)^t}$ = *sum of electrical energy produced over lifetime*

Where:

I = investment expenditure in year t

M = Operations and maintenance in year t

F = fuel expenditure in year t

E = electrical energy generated in year t

r = discount rate

n = Expected lifetime of the reactor

The CSIRO GenCost report uses lifetime values for the Solar PV and the Windfarm as 30 years and the nuclear reactor as 30 years.

These values are possible outliers, and median values are most likely, 60 years for Nuclear and 15 to 20 years for Solar PV and Wind.

The discount rate assumed by the CSIRO in the GenCost report is 0.0599% which is reasonable as are the other factors except the initial capital investment assumed at

$31,138 per kW for nuclear and 1526 per kW for Solar. The value for nuclear is higher than overseas values.

GenCost also claims the capacity factor for nuclear is 53% and Solar at 19%. Nuclear value is quoted by the CSIRO is much lower than the median.

If we factor median values as per the work in Canada and the USA, we obtain the following:

Capital cost capacity factor for SMR from the SMR Roadmap

		Capital ($/kW)	Capacity Factor (%)	LCOE ($/MWh)
SMR RoadMap	Low	4837	90	66
	Median	7098	90	87
	High	16487	90	117
GenCost	Low	7237	80	143
	High	16487	60	336

The discrepancies here are huge and worrying.

The biggest single factor influencing this is the assumed lifetime of the reactor as it appears in the discount rate when it used to the power of the lifetime of the reactor.

The discount rate factor, r, is 0.0599%

From the above table the sensitivity of the calculated LCOE to the lifetime of the reactor is readily apparent.

n	$(1 + r)$ to the power n
1	1.06
2	1.12
10	1.79
30	5.73
60	32.8

Also, the LCOE equation has limitations. It does not include the cost of the waste but the Roadmap SMR for nuclear does. Waste from Solar farms and wind farms is being buried in landfill. Again, the CSIRO GenCost figures and the LCOE equation do not include the cost of global warming estimated to be between US$1.7 Trillion to US$3.0 Trillion per year.

John Blakemore (1 July 2024).
PhD HonDUniv (Newc) ASNT BSc (NSW)
mascjohn@icloud.com
0414970758

5.10   Electricity Costs Quoted Overseas

OECD Electricity generating costs are quoted as follows ($/MWh):

Country	Nuclear	Coal	Gas CCGT
France	45.3	—	—
Japan	61.2	87.6	87.6
Korea	39.4	69.8	83.0 to 91.0
Slovakia	57.6	—	—
USA	43.9	75.1 to 96.3	40.7
China	49.9	70.6	81.5
Russia	27.4	—	—
India	48.2	64.7 to 94.6	—

All the above values for nuclear power are less than that for solar PV.

An unbiased review of the energy options is needed to confirm my findings, but at this stage I believe the Australian public are not getting an accurate picture of the cost of the renewable experiment and the role of nuclear power.

Perhaps the only way to get an accurate picture is to engage some overseas experts to study the situation.

The right decisions must be made now as the path to nuclear is going to be long and arduous.

Chapter 6

Benefits and Possible Dangers of Nuclear Energy

Summary

Generally, the most important fears regarding nuclear power are radiation exposure, nuclear meltdown and terrorism and nuclear proliferation. Radiation exposure is not a problem. Nuclear submariners receive half the dose of people standing in most cities in the world. Modern nuclear reactors are failsafe and so meltdown is not an issue. There is a metastable situation regarding nuclear proliferation. The nuclear accidents that have occurred have been due to operational error and a Tsunami. Seventy years of experience and research and development has resulted in significantly advanced nuclear technology As a result, nuclear energy is safe.

6.1 Radiation Exposure

6.1.1 Nuclear Power and Irradiation

Radiation over the centuries has been responsible for the wonderful and extraordinary variety of life on our planet.

Is Nuclear Power the Answer?
John Blakemore
Copyright © 2025 Jenny Stanford Publishing Pte. Ltd.
ISBN 978-981-5129-69-4 (Hardcover), 978-1-003-63657-1 (eBook)
www.jennystanford.com

Just as radiation can cause cancer it can also destroy cancer.

6.2 OPAL

The Open Pool Australian (Light Water) Reactor (OPAL), at Lucas Heights in Sydney, was built with the major aim of supplying isotopes for medicine and research. It replaced the HIFAR reactor, which was built in 1952. OPAL is a 20 MW reactor, and it was opened in 2007. It operates 300 days per year. There was a lot of opposition to its construction and many protests by environmental groups. In most cases, these groups were badly informed.

So, radiation exposure can be a two-edged sword. Too much can cause disease, while targeted radiation therapy can destroy cancer cells and save lives.

The claimed levels of radiation exposure from various sources are given in the following table:

Radiation exposure

Exposure	Dose
Whole-body CT scan	1000 mrem per exam
Mammogram	42 mrem per exam
Cosmic radiation at sea level	30 mrem per year
Terrestrial radioactivity	21 mrem per year
Chest x-ray	10 mrem per exam
Living near nuclear power station	Less than 1 mrem per year
Navy personnel in nuclear submarine	Approximately half-terrestrial
Whole-body exposure to death	500,000 mrem
No measurable effect	5000 mrem

Figure 6.1 The OPAL reactor at Lucas Heights in Sydney.

It is clear from the table that radiation exposure from a nuclear reactor is of no great concern. In fact, submariners in a nuclear-powered submarine with nuclear weapons receive half the normal radiation dose received by normal people living in a normal city like Sydney.

The primary unit of measurement of radiation exposure in the above table is the rem (roentgen equivalent man), while most of the measurements above are in millirem, i.e. 1000th of a rem.

6.3   Other Measurements of Radiation Dose

- 1 gray = 1 Joule/kg
- 1 rad = 10 milligray (mGy)
- 1 sievert = 100 rem
- 1 becquerel = 1 count per second
- 1 curie = 37 Gigabecquerels

6.4   Medical Uses of Nuclear Radiation

Radiation therapy is used worldwide for the treatment of cancer. Australia services Asia and the Pacific with isotopes for this technique.

LINAC

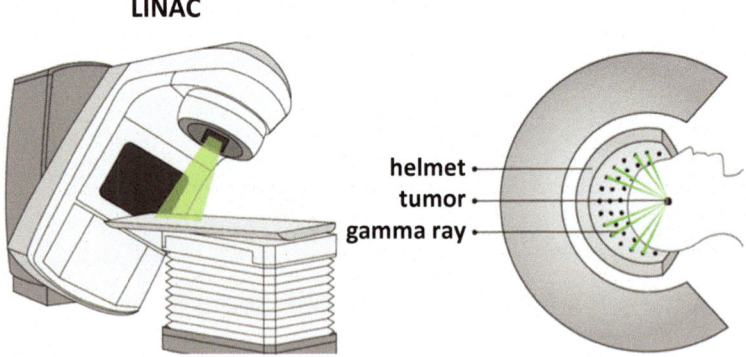

Figure 6.2 Radiation therapy used to treat cancer.

6.5   Nuclear Accidents

6.5.1   Windscale UK October 10 (1957)

The Unit 1 of Windscale core caught fire and melted. This released a significant of radiation into the atmosphere.

In this case, the reactor caught fire at a time when operators were trying to release the built-up energy. An estimated 11 tonnes of uranium ignited. Operators tried to extinguish the fire with CO_2 but that failed. Therefore, then they tried water and succeeded. Radiation escaped and reached as far as Europe. More than 200 cancer deaths are attributed to this accident.

6.5.2   Sodium Reactor Experiment (1959)

Los Angeles, California, USA, July (1959): Overheating due to a blockage in the cooling flow caused several fuel rod assemblies to reach temperatures up to 370°C and fission products were released into the atmosphere.

6.5.3   SL-1 Idaho Falls (January 3, 1961)

The withdrawal of a single control rod caused a catastrophic power surge and steam explosion. This killed all the workers on duty at the time.

Operators were in the process of attaching drive mechanisms to the control rods they had disconnected earlier for experiments. They lifted the control rods 50 cm instead of 9 cm as required. This error caused the reactor to go critical. All three workers received lethal doses of radiation.

6.5.4   Enrico Fermi Unit 1 (1966)

Frenchtown, Michigan, USA, October 5 (1966): Coolant flow blockage in two fuel channels led to the partial meltdown of two fuel assemblies at Fermi Unit 1. Fermi 1 was the nation's first and only commercially operating liquid metal fast breeder reactor. Vibrations caused coolant to be blocked.

6.5.5   Three Mile (1979)

Middletown, Pennsylvania, USA, March 28, 1979:

This resulted in only a very small radioactive release. A human operated relief valve in the primary system was stuck open and this allowed a large amount of coolant to escape. This again was operator error.

6.5.6   Chernobyl Reactor Accident (1986)

This was a direct result of a flawed reactor design and a reactor with inadequately trained personnel. Two plant workers died as a result of the explosion and a further 28 died later due to radiation exposure. 350,000 people were evacuated from the area around the reactor as a result of the accident. Large areas of Belarus, Ukraine and Russia were contaminated. Major safety mechanisms had been disconnected prior to the conduct of the experiments being done at the time.

A very wide range of changes have been made to reactor design and control in both Russia and Ukraine and the Western world as a result of this accident.

6.5.7   Fukushima Dai-ichi (2011)

The earthquake and tsunami that struck Japan on March 11, 2011, caused serious damage to the Fukushima Dai-ichi reactor nuclear power plant. The tsunami was more than twice as high as the reactor was designed to withstand. The fuel in three of the reactor cores melted. Nearly 500,000 people had to be evacuated.

6.6 Nuclear Disarmament

Nuclear disarmament is the act of reducing or eliminating nuclear weapons. Its end state can also be a nuclear-weapons-free world, in which nuclear weapons are completely eliminated. The term denuclearization is also used to describe the process leading to complete nuclear disarmament.

Disarmament and non-proliferation treaties have been agreed upon because of the extreme danger intrinsic to the possession of nuclear weapons war occurring, especially considering accidents or retaliatory strikes from false alarms. Critics of nuclear disarmament say that it would undermine deterrence and make conventional wars more common.

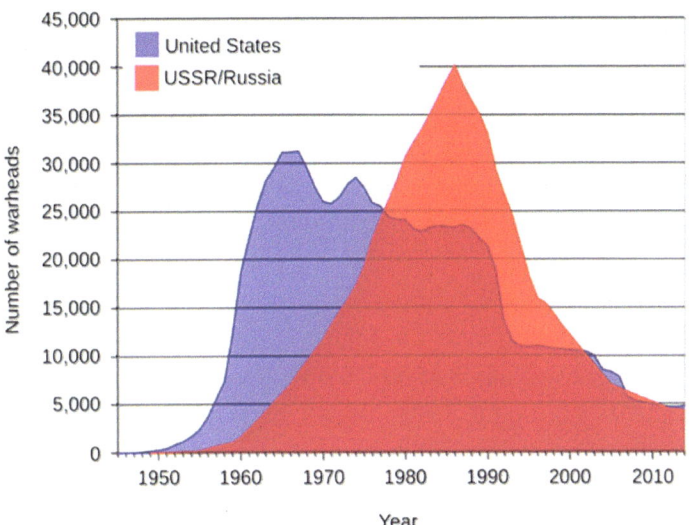

Figure 6.3 Disarmament. Reduction in nuclear weapon stockpiles USA and Russia, 1950–2010.

Organizations Aimed at Nuclear Disarmament

Nuclear disarmament groups are widespread in the world. There have been many large anti-nuclear demonstrations and protests in many countries. On 12 June, 1982, 1 million people demonstrated in New York City's Central Park.

In recent years, some US elder statesmen have also advocated nuclear disarmament, these include Henry Kissinger, and George Shultz.

In 1945 in the New Mexico desert, American scientists conducted "Trinity", the first nuclear weapons test. This marked the beginning of the atomic age. Even before the Trinity test, national leaders debated the impact of nuclear weapons on domestic and foreign policy. Also involved in the debate about nuclear weapons policy was the scientific community, through professional associations such as the Federation of Atomic Scientists.

On August 6, 1945, towards the end of World War II, the "Little Boy" device was detonated over the Japanese city of Hiroshima, exploding with a yield equivalent to 12,500 tonnes of TNT. The blast and thermal wave of the bomb destroyed nearly 50,000 buildings (including the Japanese headquarters of the 2nd General Army and Fifth division and killed 70,000–80,000 people outright, with total deaths being around 90,000–146,000.

Later, the "Fat Man" device exploded over the Japanese city of Nagasaki 3 days later on August 9, 1945, destroying 60% of the city and killing 35,000–40,000 people outright, and up to 40,000 additional deaths might have occurred over some time after that. Subsequently, the world's nuclear weapons stockpiles grew.

In 1946, the Truman administration commissioned the Acheson-Lilienthal report, which proposed the international control of the nuclear fuel cycle, revealing atomic energy technology to the USSR, and the decommissioning of all existing nuclear weapons via the United Nations Atomic Energy Commission (UNAEC).

With key modifications, the report became US policy. In the form of the Baruch Plan, which was presented to the UNAEC during its first meeting in June 1946. As Cold War tensions emerged, it became clear that Stalin wanted to develop his own atomic bomb and that the United States insisted on an enforcement regime that would have overridden the UN Security Council veto. This soon led to deadlock in the UNAEC.

Operation Crossroads was a series of nuclear weapons tests conducted by the United States at Bikini Atoll in the Pacific Ocean in 1946. Its purpose was to test the effect of nuclear weapons on naval ships. Pressure to cancel Operation Crossroads came from scientists and diplomats. Manhattan Project scientists argued that further nuclear testing was unnecessary and environmentally dangerous. A Los Alamos study warned "the water near a recent surface explosion will be a 'witch's brew' of radioactivity". To prepare the atoll for the nuclear tests, Bikini's native residents were evicted from their homes and resettled on smaller, uninhabited islands where they were unable to sustain themselves.

Figure 6.4 Nuclear bomb.

Chapter 7

Conclusion

During my travels and discussions with many eminent scientist and engineers and politicians and members of the public both here and overseas, the conversation over a meal often turns to a debate on what is the future going to be like and where will Australia be in it. In 2005, our PM John Howard was awarded the statesman of the year award. Australia was regarded as a leader internationally and greatly admired. Sadly, this is not the case today.

Overseas people are always very keen to understand the Australian culture, and often, if they are Westerners, they try to understand the ambiguity of Australian society. The ambiguity of Australian soldiers being killed in many conflicts overseas such as World Wars I and II, battles such as Gallipoli or in South Korea and more recently Afghanistan. They invariably admire our sporting success but never our industrial or innovative achievements.

To most overseas people, we seem to lack a belief in self and lean too heavily on our British heritage. We are still resource-rich but now policy-poor.

Australia had an outstanding reputation for leadership from 1900 to perhaps 1990, but since then we have become a

Is Nuclear Power the Answer?
John Blakemore
Copyright © 2025 Jenny Stanford Publishing Pte. Ltd.
ISBN 978-981-5129-69-4 (Hardcover), 978-1-003-63657-1 (eBook)
www.jennystanford.com

vanilla society on the world stage. This attitude has resulted in the demolition of our manufacturing industry. In the period from 1945 to 1960, we could make cars, planes, white goods, in fact anything. We also had a successful steel and aluminium industry, but in 2025?

As a multicultural society, we should be more independent. Many people also feel that we do not truly understand the opportunities in front of us or the opportunities presented to us in the past.

Despite this, Australia is a very balanced and friendly society and in many cases the envy of the world. However, with our calm and serenity, we have not produced a Google, a Microsoft, an Apple, an Amazon or a Paypal or a Honda, a Toyota or a Tesla.

Does this really matter? I think it does.

More to the point, Australians live in the very oldest land heavily denuded over time with what is probably the oldest existing and somewhat preserved Aboriginal culture in the world—a culture that was able to master the unique land of droughts and flooding rains and extreme temperatures of their environment.

It is indeed a tragedy that the English invaders did not appreciate the wonderful aboriginal culture and specifically how it was adapted to the Australian environment.

It is therefore probably most appropriate to examine and start an analysis of this type from the standpoint of the known history of the development of aboriginal society in Australia.

It is also important for the reader to understand my limitations and what I hope to achieve with this book.

Firstly, I was born to Australian-born parents in 1939 in the steel and coal city of Newcastle, our energy research capital. I married a Newcastle-born beautiful woman in 1965. She was 6 years my junior. I was educated in science, engineering, mathematics and nuclear physics. My beautiful wife, Deirdre, passed away from ovarian cancer in 2014.

I believe my perspective is one of a curious mind with a strong interest in improving the country I love dearly.

To do this, I had to understand the past and appreciate as much as possible aboriginal culture and my Scottish heritage.

As an engineer and nuclear physicist, I am still puzzled by the strong feelings against technological opportunities offered to us by wonderful and abundant resources. Rich coal fields in New South Wales and Queensland provided an easy pathway to riches as the world needed power and fossil fuels were an easy solution. As the world became aware of the massive damage to our environment and health resulting from global warming, the world has turned against fossil fuels as the best source of energy production because they produce too much carbon dioxide.

Unfortunately, Australia did not see the opportunity this created for nuclear energy production from our vast uranium resources. The reasons for this are readily understood—the fear of nuclear war, nuclear accidents such as Chernobyl, damage to health caused by radiation, and radioactive waste longevity, and of course the massive risks involved if there was another World War and nuclear weapons were used. Three times already we have been lucky to avoid a nuclear war (KE007 commercial flight shot down by Russia, the Bay of Pigs invasion in Cuba, and a failed computer chip in a Russian computer).

Even though 30 countries of the 195 in the world have adopted progressive nuclear policies, Australia remains drowned in old thinking.

In 1969, Australia was planning to build a 500 MW nuclear power station at Jervis Bay. Tenders were called and there were four reactors on the short list. These were two enriched uranium reactors—the pressurised water reactor (PWR) and the boiling light water reactor (BLW)—and two natural uranium reactors, the steam-generating heavy water reactor (SGHWR) and the Canadian deuterium natural uranium reactor (CANDU).

This proposal fell over ostensibly because of the cost of building the nuclear reactor and the projected cost of electricity.

At the time, Australia had numerous coal-fired stations sitting on top of coal seams like Bayswater. No consideration or thought was given to the spin-off that would occur if Australia became a more advanced nation technologically and scientifically.

I had spent most of my post-graduate career pursuing the goal of participating in the scientific revolution that I thought would follow a new technological path forged by nuclear advances.

The fundamental lack of confidence and risk taking in Australian society has always been a problem. To me, if I had not been courageous with my eyesight, I would have been blind today like my mother. Numerous of my inventions have been exploited overseas, notably in the USA, but not in Australia. The reason given to me mostly was:

"Sorry John, we can't do that, no one else in the world is doing that".

Hence, a lot of opportunities in engineering and manufacturing were not exploited. The notable exception was Colorbond, with its ground-breaking corrosion-resistant zinc-aluminium coating. The original idea was American not Australian, but we successfully exploited it. I was chief research engineer tasked with commercialising it.

It was this closed mind that led me to set up my own consulting company in engineering. With the help of others, we exploited aluminium filtration, which led to the successful aluminium can, the first Quality System in Australia—8 years before the ISO system, an improved method of strengthening steel following the dislocation carbon methods I developed for my PhD, a new operation for angle-closure glaucoma, revised algorithms for a screening device for cervical cancer, streamlined lean integration methods for training the US Air Force, development of new algorithms linking inventory to production with less working capital, and new innovative techniques for creative development of new products and processes.

Australia has always been regarded as the land down under and in many cases forgotten. One only has to remember the attitude of Winston Churchill and the way he sent our young Australian men to be slaughtered at Gallipoli and later his reluctance to release Australian ships to defend Australia against the Japanese after the fall of Singapore, to get an idea of how we were generally perceived by the political hierarchy in England. This probably explains why we get so much enjoyment out of beating them at cricket. They invented the game, but we showed them how to play it. Everything has improved since then, but our relative isolation still poses problems only partially improved with the digital revolution.

Our research and development as a percentage of GDP is low by developed nations' standards and is mainly funded by the government. Our premier research establishment, the CSIRO, has a continuous conflict of interest as the government of the day controls the funds. There are no significant other research establishments in Australia, as there are in the USA. Many times in my career after my company successfully completed an assignment in Australia, the American owners would offer me a job in the USA in research but my fear of guns alienated me from American Society. It was always a massive debate for me, pursue my technological ambitions or remain in Australia.

Australia is now once again afforded a unique opportunity now to lift its technological performance by considering nuclear energy as a reliable option or addition to renewables to produce electricity.

Don Dunstan, when he was premier of South Australia, saw opportunities for technological progress in Australia if we were supported primarily by the Japanese and the USA. Dunstan proposed that a Multifunction Polis be built in Adelaide supported by Japan to expand the education and training of Australians in advanced technology and manufacturing. This was vehemently opposed by the Trade Union movement and the idea evaporated. It is still a good idea today.

To the surprise of many, we do have a small nuclear industry at Lucas Heights in Sydney which is the home of our only nuclear reactor the OPAL. This reactor is the successor to HIFAR (High Flux Australian Reactor). These reactors were established to produce isotopes for medicine and industry.

The high cost of nuclear power generation is cited most often as the reason Australia should reject nuclear power. The method of calculating the cost of electricity, LCOE, has many variables and, of course, the final outcome is greatly affected by these. The most important ones are the following:

- Investment expenditure
- Operations and maintenance
- Fuel expenditure
- Electrical energy generated
- Expected lifetime.
- Discount rate

The variable that has the greatest impact on the comparisons that are drawn between nuclear and renewables is the plant lifetime.

There is no consideration here given to the cost of global warming or the cost of waste disposal. Waste disposal costs are included for nuclear because of its known dangers of radioactivity, but the total waste disposal costs are generally downplayed for renewables as is the cost of the possible loss of arable land.

Capital cost 1000 MW		Lifetime (Years)		Capital cost (60 years)	
Nuclear	**Solar**	**Nuclear**	**Solar**	**Nuclear**	**Solar**
US$8.0 billion	US$1.2 billion	60 years	20 years	US$8.0 billion	US$4.8 billion

As can be readily seen, if we look at the total capital cost over the probable lifetime of the reactors, the total capital difference diminishes somewhat. The above approximate calculation does not account for the diminishing value of the dollar due to inflation in the future and therefore the figures quoted for solar over the projected 60-year lifetime would be much more, approximately US$6.0 billion.

In addition, it has been reported that the cheapest power in the world is being produced by the new Finnish nuclear reactor, French nuclear reactors and possibly the new reactors in the UAE.

At the same time, to confuse matters further, Germany is closing all its nuclear reactors and the Japanese and Chinese are building new ones. Germany is now starting to understand its mistake as it is stifling manufacturing and industrial development.

So, there are many paradoxes in the complex nuclear world, but some things are certain, such as the following:

- The world is heating up.
- A nuclear reactor is expensive mainly because of very strict regulatory and safety requirements.
- Nuclear reactors can supply more reliable and predictable power than renewables.
- Waste from nuclear reactors is well controlled and may be a source of energy in the future.
- Wind turbine blades and PV solar panels are being buried in landfills at the end of their useful life, which may be as short as 15 to 20 years.
- Dirt on solar panels can greatly affect efficiency.
- Nuclear reactors do make plutonium which can be used to make a nuclear bomb.
- Australia's infrastructure is so poor that it would probably take longer to build a 1000 MW (1 GW) nuclear reactor here than most other developed countries.

- Australia does not have the immediate expertise to build a nuclear reactor or even service the AUKUS nuclear submarines, but it does have the OPAL reactor and the associated systems for medicine and research.
- We need a nuclear reactor to produce isotopes to fight cancer and as diagnostic tools for disease control, and for this the OPAL reactor, like HIFAR before it, is well regarded internationally.
- The latest designs of nuclear reactors cannot have a meltdown, they are failsafe.
- There is a significant growth in the gross domestic product as a spin-off from nuclear power as shown in Fig. 7.1.

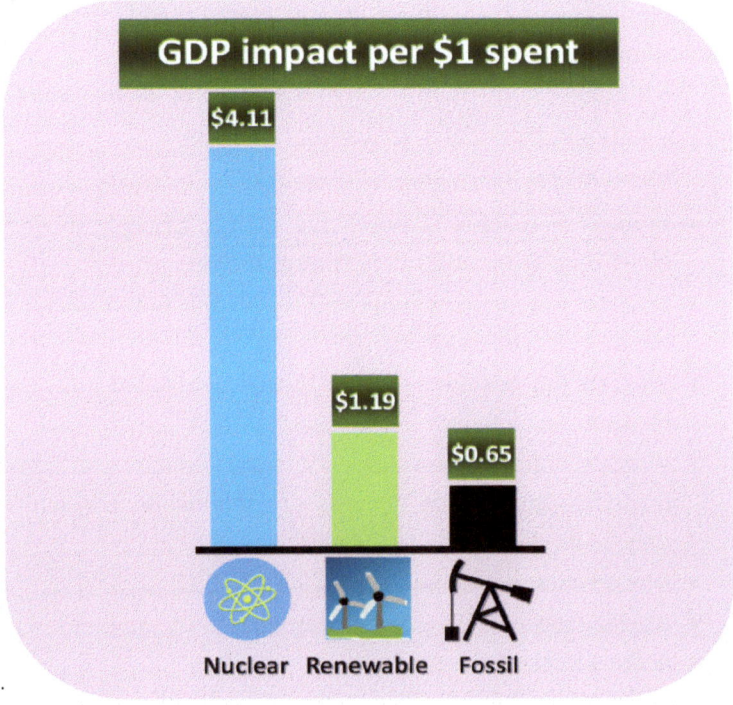

Figure 7.1 GDP growth from nuclear power. Data from Batini et al. (2022).

More and more countries are recognising that they need nuclear power, as it not only helps to ameliorate the effects of global warming, since the CO_2 emitted during operation is nil, it is also cheaper than many commentators claim and much more reliable. Wind and solar and wave all rely on the weather, while nuclear energy does not. One could argue that it was the weather, the TSUNAMI, that destroyed Fukushima. This 50-year-old Japanese reactor like all of Japan is on a fault line. The Japanese need energy and have no natural energy-producing resources. They are now pinning much of their hope for their future of energy on Perovskite solar cells.

Recent studies have produced the following relationship between CO_2 levels and the economy in France compared with Germany. France is the bottom line.

Finland has recently commissioned a nuclear reactor which supplies some of the cheapest energy in Europe. The cost of power in Germany is 41 c/kWh while in France it is 23 c/kWh, approximately half. These costs do not include the cost of global warming estimated to be US$1.3 trillion to US$3.0 trillion per year. If this cost is included, then the nuclear option is even more attractive financially.

Using the 1000 MW (1 GW) power station as an example, a major issue continuously ignored by many is that nuclear reactors last from 50 to 70 years, while windfarms and solar farms last only 15 to 20 years, although people are claiming longer. So, there could bea lifetime factor difference of three times. The capital outlay for a solar or wind farm is $1.2 billion and multiplied by 3 this is now becoming a capital outlay of approximately $3.6 billion. The estimated cost of building a 1000 MW nuclear reactor is probably $8.0 billion, but in Australia it will probably be more than that because we have dismantled much of our manufacturing industry over the past 30 years.

When drawing these comparisons, a cost has to be added in the total supply chain for the renewables to cover waste disposal and extra poles and wires. The nuclear reactor can be built on the same site as existing fossil fuel power stations, so there are minimum to no extra costs if, say, an SMR is retrofitted to an existing fossil fuel coal station.

Now, if we take into account the cost of global warming for the renewables and a nuclear reactor over say a 70-year lifetime, the nuclear option becomes even more attractive as the cost of global warming is conservatively estimated at US$1.3 trillion per year.

At the time of writing, the CSIRO has produced new data that claims that renewables reduce the cost of electricity. They look at a series of parameters that favour renewables and place nuclear at a disadvantage.

The CSIRO has suggested that the lower costs of building a nuclear reactor are only applicable in the countries which are not democratic and have low labour costs and less regulation, countries like the United Arab Republic (UAR) and South Korea. They leave out France from their analysis. Certainly, there is a redundancy factor of 5 for nuclear reactors, but if the CSIRO is going to use parameters such as a 30-year lifetime for solar cells in its analysis, then the nuclear industry can say that the latest nuclear reactors can't have a meltdown, so a lot of the costly regulations and redundancy factors can be eased. Therefore, the problem for Australia is complex and requires a more detailed analysis than I have provided here, but it must be done.

Artificial Intelligence (AI) will change the world dramatically. To use it effectively, large data centres will be needed. Despite recent significant advances with perovskites cells for solar, in the future, the enormous energy requirements needed by AI can only be satisfied by nuclear, especially when you consider the huge extra demands it will place on providing enriched uranium and maybe plutonium to feed these enrichment processes as fuel for the nuclear power stations.

Australia needs to start building the nuclear infrastructure and power stations now or it will be left even further behind in the future digital quantum world.

Artificial Intelligence is already a major driving force of the future. It needs huge data banks and storage facilities. Renewables cannot satisfy the huge potential demands for energy, but nuclear can.

Bibliography and Resources

Australian Nuclear Association. 2024. Submission to AEMO and CSIRO. www.nuclearaustralia.org.au

Batini N, et al. 2022. *Ecological Economics*. Elsevier 193, 107305.

Biegler T. 2009. Hidden Costs of Electricity. ATSe Report.

Blakemore J. 2024. Letter to Paul Graham, CSIRO. LCOE Outlier Data.

Blakemore J. 2024. Australia's Energy Needs. Summary.

Blakemore J. 2024. Cost of Nuclear Power. Summary.

Blakemore J. 2024. Wind Farms. Summary.

Bowyer JJ, Edis T. 2024. Nuclear in Australia. Inst Ec Finance.

Butcher N. 2018. SRM Roadmap. Canadian Econ & Fin Group.

Caldicott H. 2006. *Nuclear Power Is Not the Answer*. Melbourne University Press.

CSIRO GenCost Report. 2024. Review of energy costs. Full report.

Cowley S. 2024. Email to Princeton Plasma Laboratory. Princeton, USA.

Diesendorf M. 2007. *Green House Solutions*. UNSW Press.

Finkel A. 2024. theguardian.com/commentisfree. Letter, March 2024.

Glasstone S, Sesonske A. 1967. *Nuclear Reactor Engineering*. D. Van Nostrand Company Inc.

Gramling C. 2019. Global Warming Unlike Last 200 Years. General data.

Heard B. 2020. Bright New World. Sub 74 Nuclear Prohibition.

Heard B. 2021. Bright New World. Sub to CSIRO GenCost.

Joshua A. 2022. Mackinac Centre for Public Policy. Blog post. mcpp@msckinac.otrg.

Lazard. 2021. Lazard's Levelized Cost of Electricity. Wikipedia.

LCOE Costs. 2024. Cost of Electricity. Std Method Calc Gen Costs.

Lomborg B. 1998. *The Skeptical Environmentalist*. Cambridge University Press.

Maddox et al. 2009. Cost of Energy Production. Costs of All Energy Production.

Minerals Council of Australia. 2022. SMR Roadmap. minerals.org.au.

Minerals Council of Australia. 2024. GE Hitachi SMR. Summary.

Minerals Council of Australia. 2024. Terrestrial Energy. Summary.

Minnesota State University. 2024. Radiation Units. Summary.

National Council on Radiation Protection & Measurements. 2009. Exposure to Radiation Measurement. www.epa.gov/radiation.

Nuclear Accidents. 2013. Nuclear Reactors. Union of Concerned Scientists.

Office of Nuclear Energy. 2023. How Does a Nuclear Power Reactor Work?

Pittock AB. 2005. *Climate Change Turning Up the Heat*. CSIRO Publishing.

Plimer I. 2009. *Heaven+Earth*. Connor Court Publishing.

Potter B. 2023. Cost of Construction of Nuclear Power Plants. Broad details.

Rennie R. 2003. *Dictionary of Atomic and Nuclear Physics*. CXheckmarks Books.

Union of Concerned Scientists USA. 2013. History of Nuclear Accidents. Summary.

US Department of Energy. 2023. Pathways to Commercial Liftoff, Advanced Nuclear Update. Summary.

US Department of Health and Environmental Protection Agency. 2024. Radiation Exposure. Summary.

Von Weizsäcker E, Lovins AB, Lovins LH. 1997. *Factor 4*. Allen & Unwin.

World Nuclear Association. 2023. World Nuclear Association. Summary.

World Nuclear Association. 2022. Radioactive Waste Management. Summary.

Yip A, Nordhaus T. 2016. Capital Costs of Nuclear Power Stations. Summary.

Index